AF878517

NATIONAL AGRICULTURAL BIOTECHNOLOGY COUNCIL

1996

National Agricultural Biotechnology Council Report

NABC Report 8

Agricultural Biotechnology: Novel Products and New Partnerships

Edited by Ralph W.F. Hardy and Jane Baker Segelken

Published by the
National Agricultural Biotechnology Council
Ithaca, New York 14853

NABC Report 8

Agricultural Biotechnology: Novel Products and New Partnerships

The National Agricultural Biotechnology Council provides an open forum for the discussion of issues related to the impact of biotechnology on agriculture. The views presented and positions taken by individual participants in this report are their own and do not necessarily reflect the views or policies of the NABC.

Cover captions; clockwise from upper left:

Future transgenic bananas may provide edible vaccines as replacements for injected ones. Photo provided by Charles J. Arntzen, Boyce Thompson Institute.
Transgenic sheep are being developed to produce pharmaceutical protein. Photo provided by Julian Cooper, PPL Therapeutics, Inc.
A toxic metal site is phytoremediated with plants that take up metal. Photo provided by Burt Ensley, Phytotech, Inc.
A transit bus painted to show soybean seeds that provide the bio-diesel fuel for low emmission operation. Photo provided by the National Biodiesel Board.
Center: Transgenic cotton plants produce a polyester blend potentially eliminating the need for synthetics. Photo provided by Kenneth A. Barton, Agracetus, Inc.

Additional copies of NABC Report 8 are available for $5.
Please make checks or purchase orders payable to:
NABC / BTI
419 Boyce Thompson Institute
Tower Road
Ithaca, NY 14853

ISBN: 0–9630907–6–3
Library of Congress Catalog Card Number 95-071408

Printed on recycled paper

National Agricultural Biotechnology Council

Providing an open forum
for exploring issues in
agricultural biotechnology

The NABC is a consortium of not-for-profit agricultural research, extension and educational institutions established in 1988.

Member Institutions

Association of Research Directors, Inc.
Boyce Thompson Institute
Clemson University
Cornell University
Iowa State University
Michigan State University
North Carolina State University
The Ohio State University
Oklahoma State University
Oregon State University
The Pennsylvania State University
Purdue University
Rutgers University
Texas A&M University
University of Arizona
University of California System
University of Florida
University of Georgia
University of Guelph
University of Illinois at Urbana-Champaign
University of Minnesota
University of Missouri-Columbia
University of Nebraska-Lincoln
University of Saskatchewan
University of Wisconsin-Madison

Other NABC Reports Available:

NABC Report 3, *Agricultural Biotechnology at the Crossroads: Biological, Social and Institutional Concerns* (1991)

NABC Report 4, *Animal Biotechnology: Opportunities & Challenges* (1992)

NABC Report 5, *Agricultural Biotechnology: A Public Conversation About Risk* (1993)

NABC Report 6, *Agricultural Biotechnology & The Public Good* (1994)

NABC Report 7, *Genes for the Future: Discovery, Ownership, Access* (1995)

ACKNOWLEDGMENTS

The NABC recognizes the hard work and careful planning by NABC8 host institution Rutgers, the State University of New Jersey, especially the meeting coordinators Peter Day and Laura Meagher, and Jill Braun, Missy Marciante, and Joseph Robles, who collectively handled many of the meeting details. Thanks also to the NABC8 planning committee of Chee-kok Chin, Geetha Ghai, Paul LaChance, James Morris, Thomas Orton, Carl Pray, Rod Sharp, Judith Snow and Daniel Teich. Special thanks are also extended to the workshop cochairs and facilitators: Barbara Byrne, Stephen Decter, Dennis Doyle, Geetha Ghai, Bill Lacy, Alan Laskin, Michael Lawton, Jeanne Mroczko, Lynn Slifer, Judith Snow, Rick Walter, and Lynn Watkins.

Particular recognition goes to Monica Voionmaa, who provided copyediting and production assistance, and to David Poland for his thoughtful editing and proofreading.

Our sincere thanks to Roger L. Mitchell, University of Missouri-Columbia, NABC Chair 1995-96, for his leadership, support and guidance.

Ralph W.F. Hardy
NABC President

Jane Baker Segelken
NABC Executive Coordinator

CONTENTS

NABC 8: An Overview

PETER R. DAY AND LAURA MEAGHER
Co-Chairs, NABC 8 Planning Committee
Rutgers, The State University of New Jersey
New Brunswick, N.J.

A major attraction of biotechnology for investors has been the allure of new market opportunities and the prospect of revolutionary medical, food, and other products that will change our lives. However, biotechnology's critics have long expressed concern that the unpredictability of the effects of such changes are reasons for prudence and caution. In spite of some false starts, unrealistic expectations, and unfulfilled promises, recombinant DNA biotechnology is now maturing as an important discipline that will underpin much of our biological research and development during the next century. Surprising to some is the important role of agricultural biotechnology to not only the food and feed industry, but also to the chemical, pharmaceutical, environmental, and energy industries, as new products are emerging in these marketplaces.

Agricultural biotechnology is beginning to act as a matchmaker for some unexpected marriages between sectors. The National Agricultural Biotechnology Council's (NABC) eighth annual meeting — Agricultural Biotechnology: Novel Products and New Partnerships — held at Rutgers, The State University of New Jersey, New Brunswick, New Jersey on June 4–7, 1996, debated the social, ethical, economic, research, development, and commercialization issues and opportunities that the new products of biotechnology pose for consumers, farmers, industry, public interest groups, government, and universities. Asked to be provocative and to speculate, 14 plenary speakers from the public and the food, pharmaceutical, and environmental and energy sectors set the stage for intensive workshop discussions of the social and ethical issues raised by new products and the opportunities for structural and economic changes.

Following the plenary sessions, participants in three workshops tackled a set of tough questions from the viewpoints of the food industry, the pharmaceutical industry, and the environmental and energy sector.

Plenary Session Highlights

Plenary Session I: Novel Products and New Partnerships

Kenneth Barton, Vice President for Research, Agracetus, was one of the two keynote speakers. His talk, "Biotechnology: Catalyst for Change in Agriculture," set the stage by pointing out how the application of biotechnology will help to sustain population growth and food production on the earth's finite area of cultivable land. The increased speed and great breadth of current change, along with the relevance of biotechnology to industries other than agriculture, will significantly impact progress in this field. He pointed to the current acceleration of the transition from biotechnology development to product and market development and the unprecedented scope of recent introductions to the marketplace. Barton went on to review the important role that intellectual property protection will continue to have in shaping the biotechnology industry in years to come. The speed of developments in biotechnology was illustrated by the progress made in improving the strength of cotton fibers, with a single genetic engineering step responsible for a strength increase equivalent to that achieved in 30 years of classical plant breeding.

As a counterpoint to the brave new world of biotechnology, the other keynote speaker, Rebecca Goldburg, Senior Scientist, Environmental Defense Fund, spoke of the unknown environmental impacts of biotechnology. She explained the skepticism of the environmental community over promises that fertilizers and pesticides would be replaced by crops that fix their own nitrogen and protect themselves from pests and diseases and benefit the environment. The first have not materialized and engineered resistance has yet to have a significant effect in reducing pesticide applications. She described those claims as greenwashing. In her view, bioremediation has been oversold and pollution prevention will be far more effective in dealing with the problems caused by chemical wastes. Goldburg also expressed continuing concern over the hazards that might result from gene transfer between the growing range of engineered plants and animals and their wild and cultivated relatives. She cited the many examples of escapes from aquaculture facilities, pointing to added risk of transgenic fish to wild populations. And she stressed the importance of conditions attached to recent permits issued by the EPA that can lead to the cancellation of permits. Those conditional registrations represent a step toward the kind of innovative regulations the environmental community seeks to have in place.

Together, the keynote speakers demonstrated the open sharing of diverse views that has become the hallmark of the NABC conferences.

Plenary Session II: Creating New Market Opportunities for Agricultural Biotechnology

David Evans, Executive Vice-President, DNA Plant Technology Corporation, described the commercial opportunities for tomatoes that stay firm and fresh longer (ethylene-regulated tomatoes, for example, now have a shelf-life of somewhere between 40 and 90 days.) He reviewed the seven different technologies that must be used to achieve this result, noting that all of them are governed by various intellectual property claims that limit a company's ability to commercialize new developments. He underscored the potential impact upon the industry of this increasingly complex and interwoven state of intellectual property rights, citing the challenge of balancing possible obstacles to innovation with companies' need to exert control over intellectual property.

From the pharmaceutical industry, Dianne Defuria, Director of Commercial Development, Bristol Myers-Squibb, discussed the role of industrial-scale plant cell culture technology in relieving her company of its dependence on extracting the anti-cancer drug paclitaxel from the bark of Pacific Yew, which at first was peeled by hand from the trunk and major limbs in natural stands. Interesting issues of high technology facilitating conservation of trees arise from this sort of situation.

Jeff Gain, Chair of the Board of the Alternative Agricultural Research and Commercialization (AARC) Corporation of the USDA, represented the Environmental and Energy sector and discussed industrial uses of agricultural products. He stressed the difficulties in growing industrial hemp without cyclone fences, guard towers, and search lights!

Caron Chess, Director of the Center for Environmental Communications, Rutgers University, discussed the public's interest in agricultural biotechnology and their perceptions of risk. She reviewed the importance of how information is supplied, pointing to the fallacy of the overly simplistic idea that if you give people information it will change their attitude and, in turn, their behavior. Information does not make people agree with what scientific experts might call "rational." An array of additional factors enter into their formation of opinions.

Plenary Session III: Social Issues, Regulations, and Ethics

The first speaker, Sister Miriam MacGillis, Director of Genesis Farm, presented a point of view completely opposed to biotechnology as well as other interventions by humans in natural processes. She voiced her concern about world hunger, presenting a view that the lawlessness of the global market system is responsible for the crises of modern society. Her thesis is that an obsession with genetic engineering may bring about a total undermining of the life that biotechnology is committed to redesigning.

In a provocative juxtaposition of presentations, Charles Arntzen, President/CEO, Boyce Thompson Institute, devoted his presentation to the utilization of biotechnology for fighting disease among children in lesser developed countries. He gave an account of some successful research on the expression of antigenic proteins in transgenic plants that may make it possible to raise vaccines in plants against hepatitis B, bacterial and viral diarrheal disease, and other infections. The objective of this work is to create an oral vaccine that is delivered when the transgenic food plant that expresses it is consumed. The plant of choice is the banana because its fruit is eaten raw and it is a widely accessible and acceptable food in many lesser developed countries.

Ken Evans, President of the Arizona Farm Bureau, described the application of other modern technologies and spoke of his own use of a 300 horsepower sludge injection tractor remotely controlled by a portable computer. Using sensitive biotechnology-based tests to detect and reject loads contaminated with toxic materials and disease organisms, he has applied over three million metric tons of uncontaminated municipal biosolids during the last 18 years, raising the elevation of his 22 square-mile ranch by over three inches! Clearly a believer in the proactive adoption of suitable technologies, Evans predicted that industrial and chemical feedstock production will be a major source of revenue for future farmers, who will be as comfortable on the Internet as yesterday's farmers were using a shovel. He stressed, however, that innovative farming must be done with provision for recreational land use and improved environmental management.

Martine Kraus of the Center for the Study of Law and Society, University of California, Berkeley, described the importance of regulation that assists the development of the biotechnology industry without stifling innovation. A comparison of the experience of United States and German biotechnology companies was revealing as shown by Germany's 20-fold higher regulatory costs. Whereas regulation creates a known climate for companies and assures consumer confidence, it was reassuring to note her conclusion that innovation thrives independent of the regulatory framework.

Plenary Session IV: Economic and Structural Issues

The session began with a talk by Brewster Kneen, an agricultural journalist from British Columbia, who discussed the biases and assumptions implied in the expression "food industry." Stripped of its hyperbole, Kneen reduced biotechnology to a monoculture modeled on the production line. Drawing an analogy between the safety of automobiles and of biotechnology, he highlighted our ability to ignore or eliminate what doesn't fit or what is unknown, and our preoccupation with speed and precision.

Julian Cooper, of PPL Therapeutics, spoke of the value to the pharmaceutical industry of proteins produced in transgenic animals. He used as an example the attachment of a milk gene promoter to a gene encoding a therapeutic protein. The latter is expressed in the mammary gland of the animal so that the transgenic protein can be harvested and purified from its milk with no adverse effects on the animal. Many therapeutic proteins are modified after translation from DNA in order to be therapeutically active in ways that bacterial expression systems, for example, cannot handle. A case in point is the inability of bacterial systems to add sugar residues to proteins (glycosylation). The yields of transgenic proteins from mammalian tissue culture systems are low, and the costs of establishing them are high, making production by farm animals an attractive mechanism for providing some important pharmaceuticals.

In his talk about the use of plants to remove heavy metals from contaminated soil, Burt Ensley, CEO of Phtyotech Inc., spoke about the novel blending of agriculture with the environmental industry. The best results have been obtained with selected forms of an agronomic crop, Indian Mustard, that take up and concentrate toxic metals from contaminated soils. When the plants with high concentrations of metal are harvested, the biomass of plant debris from a contaminated site is only about two percent of the original mass of contaminated soil, decreasing disposal costs while leaving cleaned topsoil in place. The harvested plants can be composted or incinerated to further concentrate the toxic metals. He showed experiments in progress on lead contaminated land in the inner city of Trenton, N.J., where the objective was to render the site safe for other uses in a way that can be perceived by the community as "natural" and "environmentally friendly."

The final plenary paper, by Marylou Garr of the Ontario Federation of Agriculture, presented biotechnology from a farmer's perspective. She described the evolution over the last three years of the Ontario Agricultural Biotechnology Committee. This group was designed to promote knowledge and understanding within the agricultural industry, to improve communication within the agricultural community and between it and society at large, to influence future research and commercialization, and to encourage assessment of and access to biotechnology products for Ontario. Because the rate of discovery of new products is far more rapid than our ability to address the issues that each raises, the committee is already performing a valuable function at the intersection of research and development with farming and the public.

In between the plenary sessions, participants joined one of three workshops. While the plenary sessions described here were designed to stimulate controversy, the real heart of the meeting lay in the dialogues arising in the workshops. In those breakout sessions, participants explored implications of Creating New Market Opportunities; Social Issues, Regulations, and Ethics; and Economic, and Structural Issues for a particular industry sector. Lively debate coupled with mutual respect ran throughout those sessions. The workshop reports begin on page 23.

Those attending the Food Industry Workshop discussed the pathways to be taken by new food products, issues of communication, regulation, and consumer concern over food, and the ways in which the production of new food crops might impact the structure of the agricultural and food industries.

Food Industry Workshop Recommendations

Recommendations to help create new market opportunities for agricultural biotechnology products include:

- *Increased Public Input and Awareness: It was agreed that most importantly, increased public input and awareness must occur. Consumers must be kept informed and provide input in areas such as directions for public research for biotechnology products and processes, regulatory progress and issues, and methods for prevention of unintended consequences resulting from use of these products.*
- *Ensure That Industry Take Clear Responsibility for Risk: Participants also recommended that some mechanism be established for industry to better internalize the risk and establish improved accountability should negative consequences occur, such as the medical device industry in the case of breast implants.*
- *Establish Competitive Research Consortia: Establishment of competitive research consortia to share and leverage knowledge and expertise was felt to be a recommendation that would be successful for the industrial development of new products and processes.*
- *Allocate Monies to Examine Social, Ethical, and Legal Issues: Several participants felt that public and private research and development should include an allocation of monies to examine the social, ethical, and legal issues surrounding the topic, as is being done in the Human Genome Project.*

In preparing for a new model of agriculture and food production, the question of what the farming model will look like in 15-20 years needs to be considered. Will it be more highly industrialized with a small number of multinational conglomerates controlling agriculture and producing the bulk of the food supply? Will the small-to medium-size farming operation survive and thrive? Participants expressed the view that farmers can best prepare for the future by:

- *Staying Informed: Farmers must educate themselves on the technologies available as well as the political and regulatory climate that exists, and keep up to date on current and anticipated impacts and opportunities.*
- *Get Involved: The voice of the farming community is becoming unified and will gain effectiveness as the numbers of those involved increases. Participation in the political and developmental process is a necessity.*

- *Consider Alternatives: If vertical integration does occur, farmers must be prepared to seek alternative ways to survive and thrive. These include teaming up to form co-ops or other partnership operations that can compete with large industry, and developing production and distribution alternatives such as community shared agriculture (CSA) and organic farming.*

The Pharmaceutical Industry Workshop explored similar issues but concentrated on the newly emerging relationship between agriculture and the industry and the special problems of dealing with emerging markets, possible consumer concerns over agriculturally based pharmaceuticals, and the economic implications of bridging the gap that has existed between agriculture and pharmaceuticals.

Pharmaceutical Industry Workshop Recommendations

The group's recommendations include:

- *Develop educational instruments to improve the level of awareness in the public, the media, the farming community, and the pharmaceutical, food processing, and retailing industries.*
- *Information materials must describe the intent, benefits, risks, and risk management mechanisms.*
- *Regulatory change should be considered to meet increasing consumer demands for access to the functional food category.*
- *NABC member institutions should promote workshops and other related activities for dialogue between the agri-food and health care communities, academics, and the public.*
- *Participation of industry representatives at NABC meetings should be encouraged.*
- *NABC member institutions should encourage graduate students and post docs to attend NABC meetings by covering their costs.*

Participants in the Environmental and Energy Workshop discussed the ramifications of both immediate and long-term examples of harnessing agricultural biotechnology to new environmental and energy-efficiency objectives. They also explored the public's willingness to change lifestyles to improve the environment and the numerous economic issues that will affect the expanded connections between agriculture and this rapidly growing sector.

Environment and Energy Industry Recommendations

Recommendations to help create new market opportunities for agricultural biotechnology products that are beneficial to society include:

- *Develop strategies that represent a balance of broad perspectives based on inputs from the major stakeholders (public interest groups, industry, government, scientific community, farmers).*
- *Encourage government to take the lead in framing an approach to developing novel products (e.g. alternative fuels to replace petroleum-based fuels) that reflects long-term considerations, taking into account implications for the economy, the environment, national security, and other relevant issues.*
- *Develop educational programs at all levels from K-12 to professional schools and continuing education — emphasizing critical thinking, systems frameworks, full cost accounting and life cycle analysis, societal trade-offs (both current and intergenerational), and communication of complex issues in science and technology.*

Recommendations to use public opinion most effectively to shape decisions about new products via biotechnology include:

- *Develop a two-way system of education and communication on public issues, involving citizens' advisory groups early in the process, and giving the public a greater sense of control over the decision-making process.*
- *Devise new mechanisms to convene together diverse sectors (e.g. farmers, environmentalists, consumers, etc.) with a neutral convener in a community-based setting*
- *Document success stories and highlight positive benefits to society on environmental and energy-related products, and organize a balanced presentation of facts and societal trade-offs.*
- *Conduct more research on how to communicate with the public and how to bridge the gap between information and attitudes and behavior.*
- *Prioritize what products should come on-line first.*

A number of measures can be taken to better prepare farmers for new opportunities in this time of transition.

- *They can better inform themselves by taking advantage of educational opportunities and by becoming computer literate to be in a position to receive electronically transmitted information and use computer-based technology.*

- *The extension system needs to be broadened to encompass an expanded client base and to include new technology and production and marketing opportunities.*
- *Farmers should join with industrialists, financial institutions, environmentalists, consumer groups, universities and government — to develop a vision for the new agriculture and set a national agenda that takes into account broad views of benefits and costs to the industry and to society.*
- *Farmers should develop cooperatives for investment in new technologies and to create new products (e.g. Ocean Spray, Ontario Federation of Agriculture); such cooperatives may have a greater chance than large corporations with consumer acceptance of new and novel products.*

Finally, the Group identified new partnerships that need to forged.

- *Agricultural cooperatives need to be formed, facilitated by government and universities, and focused on new visions and common goals.*
- *Alliances need to be formed among farmers, industry, environmentalists, and consumer groups in neutral settings provided by universities and the nonprofit sector.*
- *Universities and farmers need to develop new relationships for research and extension based on new realities and market-driven strategies.*
- *New industry/government partnerships need to be formed that are incentive-driven, that facilitate technology assessment and transfer, and that maintain access to technology and capital by small/independent farmers as well as large corporations.*

In summing up the conference, Paul Thompson, of Texas A&M University, stressed the importance of trust in achieving a positive synergy between industry and public interest organizations. For this to work, both parties must accept the fact that they each have power but that it can only be exercised with constraint. Paul underlined the role of NABC conferences in facilitating "reflection over science and its impact on our future." He projected a hopeful future for this meeting's novel products.

For readers unable to benefit directly through participation in the give-and-take of these workshops, the reports, on pages (23–44), provide an overview of the multiple facets of the issues that were discussed. In the best of worlds, these reports will serve as a springboard for wider dialogues about exciting yet challenging Novel Products and New Partnerships.

Tying It All Together

Paul B. Thompson
Texas A&M University
College Station, Texas

The Rutgers conference of the National Agricultural Biotechnology Council represents a potential turning point in public debates over the future of plant and animal applications of recombinant DNA manipulation and other techniques of biotechnology. Prior to the conference, debate over food and agricultural biotechnology was focused on three product groups: herbicide tolerant crops, animal drugs (especially BST), and crops engineered with the Bt gene. Although there are a few important exceptions, most products approved for use in the United States fall into one of those three groups. Debate over those products created an alignment of interests where agricultural input suppliers, the food industry, and commodity organizations opposed a coalition of interest groups representing food consumers, environmental and animal protectionists, and small scale farmers. Universities and government were caught in the middle. Only a few products (notably recombinant chymosin, the enzyme used in the production of cheese) escaped.

The Rutgers meeting provided an overview of the products that may come on-line from agricultural biotechnology early in the next century. The papers in this conference report describe new food products that will do far more to enhance the dimensions of quality that are important to consumers: taste, purity, and visual appeal. If the food industry can ensure the safety of biotechnology foods (arguably in place now at the Food and Drug Administration (FDA)) and protect the principle of individual consent and control (labels are arguably the answer), these new products will enjoy consumer support, rather than opposition. What is more, new food crops that emphasize disease resistance or that utilize less toxic forms of pest control (such as Bt crops) should garner support from the environmental community.

Authors for this NABC 8 report also note the potential for using crops and animals in the production of pharmaceutical and industrial products. If past history is any guide, the public will find the case for new or less expensive forms of human medicine a compelling one. New industrial products are among the most exciting possibilities, as applications of both plant and microbial biotechnology permit the production of fiber and cellulose crops that can relieve pressure on the worlds forests. The potential for liquid fuels holds out the promise of transferring a significant part of global energy consumption from a non-renewable to a renewable basis. Granted, the potential for such products must not be allowed to substitute for the strides that society has made in conservation, but there is every reason to hope that they may become a new approach as we strive for a sustainable society.

What constraints pose challenges to the emergence of these new uses for plant and animal biotechnology, and what new opportunities do these novel products portend? The papers in this report provide some insight into this two pronged question. One way of tying them all together is to recognize a synergy in both the challenges and the opportunities. A negative synergy of constraints places serious roadblocks in the path of developing these novel products, but with a few key changes that synergy could become strongly positive. Those key changes revolve around two points: trust and vision.

Challenges

The novel products discussed in the chapters of this report range from new foods that are entering the regulatory process (needing "only" regulatory approval and commercialization before becoming available on grocery shelves) to liquid fuels derived from plant biomass (that are, at present, largely a gleam in the eye of bench scientists conducting basic research). Yet, in every case, new technology will be needed to make the product a reality. In the former cases, it is scale-up technology needed for commercial production, or "soft" technology in the form of better marketing and consumer information. In the latter cases, basic problems in biochemistry, molecular biology, and process engineering are yet to be solved. The current state of knowledge is therefore a constraint that must be removed before the novel products described herein can become widely available.

Removing knowledge-based constraints will require research in biology, engineering, and social science. Research will require both human and financial capital. Although a lack of either could retard development of novel products, it seems reasonable to think that lack of money is a more serious constraint at the present time. Public sector funding for all forms of scientific research has been declining, and venture or investment capital for speculative biotechnology projects outside the area of human health has become scarce. The major agricultural biotechnology companies have herbicide tolerant or Bt crops and animal drugs to develop. It is questionable whether they will have the additional resources, or be willing to invest in the research needed, to bring novel products to market.

One reason why investors may be reluctant to fund research in these novel products is that commercialization will require infrastructure and institutions that do not exist. Think for a moment about what will be required to get industrial fiber grown in Iowa or Illinois to the pulp mills that are now located near the pine forests of Alabama or South Carolina. Note that there are no commodity organizations for growers of hemp, or for producers of pharmaceuticals utilizing plants or farm animals. Do petroleum buyers or oil company executives traverse the Dakotas in search of new sources of supply? Novel products presuppose new and untried linkages between farm-based suppliers and the eventual buyers and users of their products; these linkages will have to be invented. It is reasonable to think that in the case of pharmaceutical products, vertically integrated firms will round-up the growers or husbandrymen they need, but even so, these contracting activities represent costs that must be included in the commercialization of such novel products.

The regulatory framework is also a crucial dimension of infrastructure for novel products. The FDA evaluates food safety, as well as the safety and effectiveness of new drugs. The direct environmental risks of new crops are also reviewed by the Environmental Protection Agency (EPA) and the Animal and Plant Health Inspection Service (APHIS) at the USDA. This much of the regulatory framework is in place. But do pharmaceutical or industrial crops and animals pose new risks to the food system? Is there any chance that milk from cows genetically engineered to lactate pharmaceutical products could contaminate the milk supply? Could crops engineered for fiber or ethanol production back cross into varieties produced for food, and if so, what are the risks? These do not seem to be insurmountable problems, yet they will require forms of regulatory oversight that do not currently exist.

Developing institutions and infrastructure will be very difficult if there is public opposition to the products themselves. Vertical organization of production may be the most cost effective method for coordinating suppliers or for maintaining the quality control needed for new food, pharmaceutical, and industrial products, but such changes in the structural organization of agriculture may be opposed by consumers and farm groups alike. Because it is unclear whether people will oppose the integration and industrialization of agriculture, the uncertainty itself comes to constrain efforts to build infrastructure or to design institutions, as well as being a disincentive for investing the money needed to support additional research. If there is a residue of public resistance to agricultural biotechnology—as a result of acrimony over BST or as a reaction against "playing God," perhaps—the uncertainties are heightened, and the risks of undertaking the research and development of these novel (hence speculative) products begins to seem formidable indeed.

There is thus a synergy that binds the knowledge constraints, the funding constraints, the infrastructure constraints, and the public acceptance constraints. Each constraint tends to reinforce the other three. How could the public be anything but skeptical of products descended from the heritage of controversy over herbicide tolerant crops and BST, especially when even their boosters acknowledge that too little is known at present to make those products a reality, and when even venture capitalists are reluctant to invest in them? Yet public skepticism retards the development of infrastructure (including regulatory frameworks) and undermines the support for funding that would ultimately address public concerns. One might conclude that the likelihood of seeing any of those novel products in our lifetimes is slim.

Opportunities

But, on the other hand, those products are more attractive to consumers, to environmental interests, and, if fears of vertical integration can be assuaged, perhaps even to small-scale producers. It is certainly the case that high quality fruit and vegetable production is already a mainstay of family farmers in the Northeast. New pharmaceutical products may well be produced on scales quite commensurate with small farms, and if drug companies do not insist on controlling the entire process, those products, too, may provide opportunities for small-scale producers to maintain a way of life. As such, it need not be a foregone conclusion that the public will oppose these products. A new coalition of interest groups might provide impressive support for them.

If a new coalition forms, it will have the capacity to address problems of infrastructure and funding in novel ways. Perhaps it will be possible for environmentalists to work with producers and industry to restructure the regulatory process for novel products. Perhaps regulatory functions can even be incorporated into organizations that reflect multiple constituencies, on the model of the Underwriters Laboratory. Small- and large-scale producers may wish to form new kinds of cooperative organizations to ensure that they maintain control over the market structure for those new products, and they may find themselves in partnership with industry giants as well as public interest groups that have a stake in seeing those technologies develop. Clearly such coalitions will have advantages in attracting investment funding, if only because they reduce uncertainty. The possibility of co-ops, working with consumer and public interest groups, in partnerships with industry for development of novel products produces a mind-bending alteration of what we might mean by "community supported agriculture" in the 21st century.

The potential for such new social amalgamations is no less innovative and radical than the biological amalgamations discussed throughout this volume. Tomorrow might look very different, and not only because we will have genetically engineered foods, fuels, and pharmaceuticals! This, of course, implies greater involvement of more people in setting the agenda for research on agricultural biotechnology, and it probably means that more kinds of research will need to be done. The fact that such research will be popular bodes well for getting it supported not only by government but by private foundations, too. There is potential for a positive synergy, for research progress to beget public enthusiasm and cooperation, and for this enthusiasm to beget new institutions and new approaches to infrastructure. And these new institutions may provide both new sources of capital and renewed vigor in the old sources of government, foundation, and Wall Street fundraising.

This is an admittedly sketchy picture of the new positive synergy. A lot of problems will need to be solved along the way, and many of the real concerns about the novel products themselves, as well as residual problems from existing products of agricultural biotechnology, still need to be addressed. Nevertheless, there is reason to think that new coalitions will have enthusiasm for such tasks to the extent that they see themselves engaged in activities that are truly aimed at improving the human condition, including, of course, the environment on which that condition depends. Will it happen? Who knows, but at least two "big picture" items are key. One is trust, the other is vision.

Keys to Positive Synergy: Trust

A world in which the food industry, small and large food and fiber producers, and public interest organizations representing environmental, animal, and consumer interests work with each other (as well as with other industries) will require new levels of trust. It would be naive to suggest that such collaborations will come easily, and it is important to think carefully about what is meant by trust. Philosopher Annette Baier offers some guideposts in her recent book *Moral Prejudices: Essays on Ethics*. Writing from a feminist perspective, she encourages us to resist an interpretation of trust that confuses trust with power. It is an unavoidable fact of human society that some of us depend upon others for our jobs, our food, our health care, and for other aspects of our well-being. These relationships define relative positions of power and dependence. In democratic societies, we often speak of the rules for constraining or using power in terms of a social contract. Contractually specified power relations are "minimal moral traffic rules, designed to restrict close encounters between autonomous persons to self-chosen ones" (Baier 1994). Baier believes that while the social contract is indispensable to democracy, it is a poor model for trust. A better model emerges out of women's experiences raising children, caring for families, and maintaining networks of care in the community.

Trust applies in situations where explicitly specified rules and contractual requirements are inadequate or inappropriate. To trust another person is to expect that there will be situations in which the trustee must exercise judgment and discernment of a sort that cannot be anticipated and spelled out in terms ofcontractual provisions. Trust requires that trustees make such decisions based upon a genuine and comprehensive concern for the interests, the well-being, and the whole person of those who have placed their trust in them. Yet it is also true that relationships of trust are bounded, that they come with limits and with opportunities to monitor, revise, or revoke the trusting relationship. To put this as Baier does, children do not trust their parents; their dependency upon them is too great. Parents do trust their children to teachers, to baby-sitters, or to friends, however, and in doing so they expect something more subtle and responsive than letter-perfect adherence to the terms of an explicit contract. They expect situations that have not been anticipated by the contract, and that the trusted party will maintain trust by acting on behalf of the child, rather than interpreting the ambiguity or unexpectedness of the situation as an opportunity to revert to self-interest.

This means that some of the things Rebecca Goldberg and Caron Chess discuss—regulations that utilize conditional approval or site licenses that give citizens the power to revoke approval—may not be forms of trust in Baier's sense. Those measures seem to be negotiations of power, and our experience with such negotiations tells us that where there is ambiguity in the terms of the rule or the contract, we can expect each party to adopt the interpretation that is most favorable to themselves. Make no mistake, I am not arguing against institutions that spread power and that frame power relationships within explicit statements of rules and responsibilities. But let us not deceive ourselves into thinking that these are relationships of trust. As Baier writes, "Trust is rational in the absence of any reason to suspect in the trusted strong and operative motives which conflict with the demands of trustworthiness as the truster sees them. But trusting can continue to be rational, even when there are such unwelcome suspicions, as long as the truster is confident that in the conflict within the trusted the subversive motives will lose to the conformist motives" (Baier, 1994).

I take this to mean that each of the parties who must collaborate in order to create positive synergism for new products in biotechnology must share and remain true to a common vision, even when there are substantial individual incentives to defect from it. Perhaps an example will make this point more clearly. Many members of the National Agricultural Biotechnology Council have built a trust relationship with farmers, ranchers, and rural communities. Farmers, ranchers, and rural communities represent diverse interests, and there has always been difficulty when the knowledge produced in agricultural universities has benefited some representatives of agriculture at the expense of others. Agricultural research is also justified in terms of benefit to the larger non-agricultural public, but the trust relationship between agriculture colleges and their clientele continues to imply that the scientists, administrators, and faculty of agriculture colleges and research stations are making judgments on behalf of agriculture. Yet, though there can be little doubt that many of the novel products described in these pages are intended for public benefit, it is questionable for some (at least) as to whether they are truly agricultural biotechnology.

Novel products are exciting applications of the plant and animal sciences to the creation of products for human medicine or industrial processes, and they are in this sense good for the scientists, administrators, and faculty of agricultural colleges. But it is not obvious that such products will ever be produced by people who today recognize themselves as farmers or ranchers, or on the terms that characterize the traditional economic structure of agriculturally based communities. Are we being entirely honest, are we maintaining our trust, when we describe these as novel products of agricultural biotechnology? Will small-scale niche producers ever produce pigs for pharmaceutical production or xenografts, or will those pigs be cared for in facilities that are wholly owned and operated by medical supply or drug companies? Is it fair to say that we are helping agriculture, when in fact the most we are doing is creating wage labor opportunities to care for pigs in the employ of large multi-national corporations?

None of this is to say that such innovations should not be pursued, nor is it to say that scientists currently employed in agricultural universities should not do the work. Yet here, I think, is a way in which members of the research world can defect from the common vision of agriculture that underlies rural community support for research and education. Researchers may not *mean* to betray a trust. They may think, with some justification, that they are acting within their rights, but that is only to underscore Baier's point. Relationships of trust are not simply a matter of following the rules and obeying the law. Trust requires judicious thought about whether what is done is truly done for the sake of those who trust, or whether personal motives and opportunities may have colored the common vision. Since the point has now come to turn to vision, let us take up the second key.

KEYS TO SYNERGY: VISION

Many of the institutional members of the National Agricultural Biotechnology Council have been engaged in processes of "visioning" during 1995 and 1996, so much so that the term may have lost whatever appeal it once had. Perhaps we would be better off to use Sister Miriam MacGillis' term, "cosmology." Whatever the terminology, the idea is to recognize that we tell one another stories about where we have come from and where we are going. Although these stories (visions) are seldom made explicit, they serve as underlying assumptions about what is real and what is worth doing. One of those stories that was told for centuries goes by the name of The Enlightenment Project. This intimidating phrase signifies the vision of science and technology relieving the many dimensions of human misery: hunger, disease, and deprivation. The individuals who conceived this vision, Francis Bacon, Thomas Hobbes, Robert Boyle, and Rene Descartes among them, lived during the sixteenth century, almost five hundred years ago. Although in some respects their vision has come true, they also thought that unleashing the forces of scientific inquiry would just automatically, naturally redress the social problems of injustice, of cruelty, of jealousies, and of corruption.

No one believes that science will automatically redress those moral problems anymore. What is more, we have learned that our technological solutions to problems of resource scarcity can create new problems of their own, problems in the form of pollution, health risk, and environmental degradation. There is no doubt that we must move beyond The Enlightenment Project, that we must find a vision (a cosmology if you will) that does not leave the resolution of moral problems facing us to a blind and inexorable notion of technical progress. However it has succeeded in other dimensions, that vision of progress has failed to create a state of bliss at the end of history. We have many problems that have always been with us, and more than a few new ones, to boot.

The mechanistic vision of nature and humanity that emerged from The Enlightenment Project has proved an inadequate basis for understanding social problems, environmental impacts, and the moral significance of technology's unexpected and unwanted effects. That much is clear. The Enlightenment Project may also have encouraged habits of mind that fixated on patterns of domination: domination of women, domination by race, domination of nature. The Enlightenment Project may have made us less sensitive to the potential for adapting to and living with our limitations, rather than trying to overcome them. It does not follow, however, that we would have been better off never to have undertaken The Enlightenment Project, nor does it follow from any of this that we should abandon all use of biotechnology. There is a huge inferential leap being made by biotechnology's critics in this regard. We should indeed devote serious and critical attention to the implicit vision that we may be working with. Formulating this vision is more important than any specific application of biotechnology, but only petty jealousy would permit us to blame biotechnology for our collective failure to address the fundamental spiritual issues of our time. The practitioners of biotechnology have been spectacularly successful in acquiring the rewards of new buildings, financial success, and public attention. It is easy (perhaps inevitable) that those who call for moral reflection over our pursuit of technological advance should feel neglected and cast aside. It is only too human to wish ill upon an industry that sometimes seems willing to exploit the state of disarray and dissolution in our collective moral vision by making fantastic promises, but the simple fact is that we will *need* biotechnology in any adequate response to the ecological and moral crises of our time.

Our new vision will almost certainly be meliorist. This forty dollar bit of philosophical jargon means that we will abandon the view that any scientific, technological, or political project will just automatically solve all our problems. We will abandon the view that progress is either inevitable or automatic. This means that we abandon both The Enlightenment Project of dominating nature's mechanisms and the conceit that there ever was a truly harmonious nature (disrupted by human activity) to which we must return. In place of this, we will recognize the ecologists dictum: everything is connected to everything else.

When we "solve" one problem, we must be prepared to find and fix whatever breaks next. Life is indeed an evolutionary process, but not one that is tending toward a state of perfection or bliss. Instead, each adjustment (whether human caused or not) adumbrates through the global ecosystem, producing myriad changes, and opening niches for novel life activities (both human and not). The evolutionary process goes on and on. It is never finished.

The vision for human action is one of ameliorating intolerable situations, one after another, and using whatever tools are at our disposal to do so. Plant and animal biotechnology will be among them. Our use of those tools should be evolutionary, rather than revolutionary. We should use biotechnology as wisely as we possibly can, and we should be vigilant in monitoring and learning from our mistakes. We can be sure that there will be some unintended consequences from using biotechnology, but there will be unintended consequences from not using biotechnology, too. We should not expect magic bullets from biotechnology or from not using it. We should do what we can humbly and reflectively. That is what meliorism means.

Yet, I would like to say a few words on behalf of Francis Bacon, Thomas Hobbes, and the other gentlemen of the Enlightenment. Their mechanistic vision of nature indeed lead to the revolutions in health care, agriculture, and engineering that brought about the modern world. Would that we could formulate a vision half so productive! And while we cannot believe that exclusive attention to mechanisms will solve our moral and environmental problems, we should be reminded that thinkers like Bacon and Hobbes fully expected to spend two hours engaged in debates over philosophy and theology for every hour spent at the laboratory bench. Who knows what vision and what synergy we would discover if we were to collectively spend a similar proportion of our research effort on philosophy, history, and literature.

Reproducing a vision is even more important than the initial discovery of vision. The time that Enlightenment thinkers spent in debate over moral subjects not only honed their vision of nature and human possibility, it spread that vision to the clientele who would utilize scientific knowledge. The challenge of disseminating a vision is altogether different today. For one thing, we have tools of mass media at our disposal that were unavailable to Bacon and Hobbes, and we have the advantage of a generally literate society. But we also have a much larger client base; one that spans a much larger proportion of the population. It is far from clear whether we can afford to invest less in forming, articulating, defending, and disseminating a vision than could the founders of the Enlightenment, though it is evident that invest less we do.

The National Agricultural Biotechnology Council and its national conferences are for that reason all the more precious. They represent one of the few forums in which reflection over science and its impact on our future can take place. The novel products discussed in this meeting provoke a new reflection on what that future might entail. It is, for me, a more hopeful future than we have hitherto seen. It will not come about without more work in both the laboratory and in reasoned public debates. To that end, let us dedicate our hearts and minds.

REFERENCES

Baier, Annette C. 1994. *Moral Prejudices: Essays on Ethics*. Cambridge: Harvard University Press.

PART II

WORKSHOP REPORTS

Environment and Energy Industry

Cochairs

Barbara Byrne
Natural Science and Mathematics
Richard Stockton College

Stephen Decter
Ecopolicy Center
Rutgers University, Cook College

Jean Mroczko
N.J. Dept. of Environmental Protection

Judy Snow
AgBiotech Center
Rutgers University, Cook College

The Environmental and Energy Workshop Group considered a number of issues in the light of precedents being set by specific new products and services, as well as possible long term ramifications of new markets being opened up through partnerships between agriculture and these sectors. Participants had limited experience directly in these sectors, but perceived several generic opportunities and concerns.

The Group identified numerous opportunities for non-traditional products. High among these is the opportunity to replace non-renewable petroleum-based products with products derived from renewable plant or microbial products. The public is likely to view environmentally friendly products in a positive light, particularly when they bring about a corresponding decrease in the use of chemicals and are more sustainable than traditional crops, and if they bring about new economic growth.

However, participants perceived that the lack of a consensus vision on the future of agriculture is impeding both producer adoption and public acceptance of new products from biotechnology. A better understanding is needed of economic and societal trade-offs based on full-cost accounting of benefits and costs. Strategies were suggested to involve the public at an early stage in the decision-making process through more comprehensive communication and education initiatives.

Creating New Market Opportunities

The Environment and Energy Group identified the most promising opportunities for non-traditional products from agricultural biotechnology. These included:

- Petroleum alternatives based on renewable resources including ethanol, industrial oils from plants, rubber alternatives (guayule), plastics based on starch or microbial products, industrial chemicals replacing ethylene, starch based compostable containers for the food industry.
- A variety of products resulting from linking pharmaceuticals with plant and animal molecular biology.
- Stress tolerant and pest-resistant plants for agriculture and forestry production, and agricultural animals with more efficient metabolisms.
- Industrial fiber alternatives for the paper industry.
- Plant and microbial bioremediation techniques.
- Nonfood products developed from sludge and sludge-using crops.

A number of factors were recognized as barriers to creating new market opportunities:

- Lack of consumer understanding of science-based issues and the difficulty of communicating to non-science publics who are skeptical and fear the unknown consequences associated with new technologies.
- Reluctance of producers to move into new ventures because of an insufficient consensus vision on the future of agriculture and the role of producers in the new agriculture, uncertainty about the potential risks assumed by first adopters, and lack of understanding of the economics of the new agriculture and the potential profit in value-added products.
- Lack of an appropriate basic marketing infrastructure for non-traditional products including handling, transportation and processing through alternative marketing channels; and inadequate performance data on the marketing of new products.
- An insufficient understanding of economic and societal trade-offs based on full-cost accounting of internal and external benefits and costs such as waste disposal vs. reuse, energy inputs vs. energy outputs, land required for food vs. nonfood products etc.
- A public policy framework that may stifle rather than provide incentives to engender marketing opportunities for nontraditional, nonfood products.
- Lack of venture capital and funding for starting up some new ventures and for the commercialization and marketing of new technologies.

- Inadequate technologies and facilities to manage wastes associated with phytoremediation and to remove potential contaminants in sludge.

Key actions need to be taken to overcome barriers and to seize opportunities:

- Formation of producer and marketing cooperatives based on a consensus vision for the new agriculture, with better assurances that producers have significant decision making roles in vertically integrated enterprises with most power and profit residing elsewhere; and the development of new marketing channels for nontraditional products.
- Formation of research and development teams and technology transfer at universities with cross-cutting multidisciplinary and systems frameworks, providing assured sources of genetic materials, technical assistance, and providing neutral environments to convene stakeholder partners.
- Development of new models for funding based on equitable investments in research and development and providing funding partnership roles for institutions in the public, private and nonprofit sectors.
- Formulation of models for life cycle analysis and full cost accounting to better assure that the development and commercialization of new technologies solve existing problems without creating new ones such as potentially polluting by-products of new technologies and processes, the excessive use of prime farmland for crops used to produce new nonfood products (e.g. corn to produce ethanol)
- Identification of the key stakeholders and their common interests to facilitate the formation of partnerships.

A number of measures can best assure that the public will play an appropriate role in the creation of new market opportunities:

- Explore ways to stimulate public interest and involve publics early in the process of product development.
- Encourage the formation of community-based interest groups to advise the biotech industry.
- Encourage the nonprofit sector to develop visions for the future and to facilitate joint initiatives (such as the Turner and Kellogg Foundations are doing).
- Develop educational programs at all levels to encourage thinking in a systems framework, the identification of societal trade-offs, and full cost accounting — in the development and marketing of products based on new technologies and processes.

Recommendations were made by the Environment and Energy Group to help create new market opportunities for agricultural biotechnology products that are beneficial to society:

- *Develop strategies that represent a balance of broad perspectives based on inputs from the major stakeholders (public interest groups, industry, government, scientific community, farmers).*
- *Encourage government to take the lead in framing an approach to developing novel products (e.g. alternative fuels to replace petroleum-based fuels) that reflects long-term considerations, taking into account implications for the economy, the environment, national security, and other relevant issues.*
- *Develop educational programs at all levels from K-12 to professional schools and continuing education — emphasizing critical thinking, systems frameworks, full cost accounting and life cycle analysis, societal trade-offs (both current and intergenerational), and communication of complex issues in science and technology.*

Social Issues, Regulations and Ethics

The Environment and Energy Group considered what the public is most likely to be concerned about with new agricultural biotechnology products in this sector. The most widely cited concern is:

- Unanticipated and potentially adverse impacts on environmental quality.

Additional concerns are:

- Potential negative impacts on food, human health, and costs to consumers,
- Doubts about the trustworthiness and reliability of the biotechnology industry, particularly with respect to food products, and
- Loss of biodiversity

Features of new agricultural biotechnology products most likely to be viewed positively by the public in this sector are:

- Replacement of products based on non-renewable resources with those based on renewable resources.
- Potential for environmentally friendly products and processes, with a corresponding decrease in the use of chemicals and pesticides.
- Potential for new classes of products that would be more sustainable than traditional crops and would generate new economic growth.
- Optimism that biotechnology can provide better solutions to difficult problems.

A number of ethical issues that might arise from the introduction of new products include:

- Equity issues relating to proprietary rights and access of small farmers to new technologies and markets.
- Effects of new technologies on established social and economic structures, such as the displacement of traditional products (e.g. vanilla), or the use of land for non-food crops and the effect of this nonfood use on the global food supply.
- Intergenerational responsibility for conserving non-renewable resources.

Participants debated whether there are unnecessary obstacles posed to new agricultural products by regulation. They concluded that:

- Regulatory levels seem appropriate (and less of an obstacle than in the food and pharmaceutical industries), although paperwork can be burdensome and rules are sometimes applied too rigidly (e.g. academic labs governed by industry regulations).

There also was some concurrence in concerns about troublesome gaps in the regulatory coverage of new agricultural products:

- Lack of consistent international standards leading to unfair trade advantages,
- Uncertainty about which agencies will regulate new non-food products,
- Growing weaknesses in the regulatory staffs (cutbacks and inadequate training), and loss of objectivity of the review process with abolition of entire agencies such as OTA.
- Inability to deal with specific "bad actors" without penalizing an entire industry,
- Need for better understanding and balance between regulation/ enforcement and incentives.
- Lack of information on gene escape when engineered plants are released for large- scale production.

Several recommendations were suggested to use public opinion most effectively to shape decisions about new products via biotechnology:

- *Develop a two-way system of education and communication on public issues, involving citizens' advisory groups early in the process, and giving the public a greater sense of control over the decision-making process.*

- *Devise new mechanisms to convene together diverse sectors (e.g. farmers, environmentalists, consumers, etc.) with a neutral convener in a community-based setting.*
- *Document success stories and highlight positive benefits to society on environmental and energy-related products, and organize a balanced presentation of facts and societal trade-offs.*
- *Conduct more research on how to communicate with the public and how to bridge the gap between information and attitudes and behavior.*
- *Prioritize what products should come on-line first.*

Economic and Structural Issues

The Group identified both potentially positive and negative impacts for agriculture of new products being adopted by the marketplace in this industry. Among the most positive impacts are:

- Reduction of external costs to the environment and loss of non-renewable natural resources while offering the potential to increase productivity by lowering input costs of production.
- Expanding markets for traditional agricultural products e.g. stress tolerant agricultural plants, crops produced with reduced use of chemical fertilizers and pesticides.
- Opening new markets for alternative products to replace those which are petroleum based or which consume non-renewable natural resources e.g. industrial fibers, liquid fuels etc.
- Creating new markets for agricultural plants (e.g. phytoremediation) and creating income opportunities for farmers in sludge management (e.g. tipping fees, and marketable by-products from clean sludge).

In considering potential negative impacts on agriculture of new products, there was consensus that whether or not the impacts are positive or negative will depend on how technologies are brought forward and on whether new organizational structures are developed.

- There could be a negative impact on small, independent farms if farmers lose decision-making control in vertically integrated enterprises, if there is excessive concentration of power in large corporations, or if there is unequal access to new technologies.
- Unless the full consequences of initially introduced products are carefully thought out and publicly discussed, the increased suspicion of consumers will depress the potential marketability of new products.

- Some believe that unless land use issues are addressed in timely fashion, emphasis on nonfood/nonfeed products could decrease acreage for food agriculture or further deplete forest areas; others believe that market pricing could prove to be a sufficient counter weight to any drastic changes.

A number of measures can be taken to better prepare farmers for new opportunities in this time of transition:

- *They can better inform themselves by taking advantage of educational opportunities and by becoming computer literate to be in a position to receive electronically transmitted information and use computer-based technology.*
- *The extension system needs to be broadened to encompass an expanded client base and to include new technology and production and marketing opportunities.*
- *Farmers should join with industrialists, financial institutions, environmentalists, consumer groups, universitiesn and government — to develop a vision for the new agriculture and set a national agenda that takes into account broad views of benefits and costs to the industry and to society.*
- *Farmers should develop cooperatives for investment in new technologies and to create new products (e.g. Ocean Spray, Ontario Federation of Agriculture); such cooperatives may have a greater chance than large corporations with consumer acceptance of new and novel products.*

Finally, the Group identified new partnerships that need to forged.

- *Agricultural cooperatives need to be formed, facilitated by government and universities, and focused on new visions and common goals.*
- *Alliances need to be formed among farmers, industry, environmentalists, and consumer groups in neutral settings provided by universities and the nonprofit sector.*
- *Universities and farmers need to develop new relationships for research and extension based on new realities and market-driven strategies.*
- *New industry/government partnerships need to be formed that are incentive-driven, that facilitate technology assessment and transfer, and that maintain access to technology and capital by small/independent farmers as well as large corporations.*

Food Industry

Cochairs

Bill Lacy
Cornell Cooperative Extension
Cornell University

Michael Lawton
Rutgers University,
Cook College

Lynn Slifer
N.J. Commission on Science
and Technology

Lynne Watkins
American Cyanamid Company

Biotechnology-based products and processes will likely have a significant impact on the agricultural and food industries over the next several years. The nature of that impact is as yet unclear, and has the potential for both positive and negative outcomes. Topics discussed as part of the workshops at this year's NABC meeting attempted to provide the basis for beginning to understand the impact of increased use of biotechnology in food production.

The objective of the food industry workshop was to:

- evaluate new market opportunities for biotechnology in the food industry,
- examine social issues, regulations, and ethics as they relate to biotechnology in the food industry, and
- highlight the economic and structural impacts of new products on agriculture.

Creating New Market Opportunities

Participants were asked to identify the most promising opportunities for new nontraditional products from agricultural biotechnology. Not all participants were in agreement that biotechnology-based products presented new market opportunities, but rather some considered that they presented a threat to traditional products and methods. This concern was primarily based upon the unknown impacts of nontraditional products on the ecosystem, health & safety, and existing markets for traditional products.

Areas of opportunity were identified as (in random order):

- Increased Food Quality and Safety: Increased nutritional value, better taste, longer life, and food safety through developments such as packaging indicators. Development of the area of nutraceuticals for added nutritional/health value as well as removal of nutrient absorption properties and natural toxicants.
- Vaccine Production and Delivery: Development of agricultural strains for vaccine production and agricultural/food products for vaccine delivery.
- Increased Producer Productivity: Increasing producer productivity through higher yield crops, new geographic locations for crops (such as arid areas or marginal soils), and "pharming". Increased productivity will result in a net increase in the available food supply.
- Processing for Increased Efficiency: New and improved methods for food harvesting and processing, as well as the use of food processing by-products for other applications. This area provides great potential for improvement, given that current worldwide post harvest loss is significant.
- Environmentally Benign/Beneficial Food Products and Processes: New agricultural crop strains that will not adversely impact the environment and biodegradable/by-product packaging.

Considerable discussion about the drivers of these new areas of opportunity took place. Participants felt that past and present drivers were primarily economically motivated, e.g. producer and corporate profits. Some participants believe that food quality and nutritional value are becoming more important, while others maintained that economics will continue to be the major influence in the development of new areas of opportunity.

Numerous barriers exist in developing new market opportunities, primarily due to:

- Consumer Acceptance: The issue of trust and education repeatedly surfaced in workshop discussions. Trust represents the comfort level of the consumer with the expectation that nontraditional food products will or will not be beneficial to them. Consumers are skeptical that government and industry truly have their best interests at heart, and require concrete proof that this is the case. Education is an issue because the majority of consumers are likely not educated about biotechnology-based agricultural and food products. Because of the lack of trust and education, consumers tend not to trust nontraditional food products. Many participants felt that an effort to learn what consumers really think and their key concerns is critical to gaining acceptance of alternate food products.

- Intellectual Property Rights: Intellectual property issues create barriers to developing new market opportunities because patents are too broad, inhibiting innovation and entry of other similar developments. In addition, areas that cannot be protected are not as readily developed, such as nutraceuticals, where many substances are widely available. Industry does not want to expend dollars or human resources on development of products that cannot be effectively protected from competition, regardless of the benefit to the public.
- Complicated/Unpredictable Standards and Regulations: Lack of definition (in some areas) and cohesiveness of national and international standards and regulations governing biotechnology-based agricultural and food products creates a significant barrier to entry of new products. This lack of international agreement on standards and regulations affects the public trust, in turn slowing the market entry of nontraditional products.

Other barriers to entry include the capital costs associated with changes in infrastructure necessary for new products and processes, the shortage and lack of integration of resources available to develop and market these products and processes worldwide (especially developing countries), and the lack of a comprehensive management/integration system for transferring technologies to the field.

The ability of the industry to overcome these barriers will dictate the future success of biotechnology related food products and processes.

Key roles to be played in overcoming barriers and seizing opportunities include:

- Open discussions and involvement with consumers regarding the positive and negative implications of the use of these alternative products and processes. The availability of information and education on the topic will frame these discussions effectively. Mechanisms for assuring that industry will appropriately manage the introduction and control of these products and will deal responsibly with unintended consequences such as creating environmental imbalances. Full disclosure labeling of agricultural and food products produced using biotechnology.
- Worldwide harmonization and streamlining of regulations and standards to increase public trust and create an environment for more efficient introduction of new products and processes.
- Carefully cultivating and maintaining a proper balance between development of patentable and freely available products and processes.

Participants agreed that consumers need to recognize their own responsibility in this issue, educate themselves, and become actively involved in open dialogue and discussion. The public can best play a role if they are educated and aware of the research and development, regulatory, and political issues.

The group's recommendations include:

- *Increased Public Input and Awareness: It was agreed that most importantly, increased public input and awareness must occur. Consumers must be kept informed and provide input in areas such as directions for public research in biotechnology products and processes, regulatory progress and issues, and methods for prevention of unintended consequences resulting from use of these products.*
- *Ensure That Industry Take Clear Responsibility for Risk: Participants also recommended that some mechanism be established for industry to better internalize the risk and establish improved accountability should negative consequences occur, such as the medical device industry in the case of breast implants.*
- *Establish Competitive Research Consortia: Establishment of competitive research consortia to share and leverage knowledge and expertise was felt to be a recommendation that would be successful to the development of new products and processes.*
- *Allocate Monies to Examine Social, Ethical, and Legal Issues: Several participants felt that public and private research and development should include a component for allocation of monies to examine the social, ethical, and legal issues surrounding the topic as is being done in the Human Genome Project.*

Social Issues, Regulations and Ethics

Participants were asked to identify the most important public concerns with new agricultural biotechnology products. The following were identified as most important:

- Safety and Health: Concerns over the safety of biotechnology-based food products, the potential for allergens, and the potential for decreased nutritional value contribute to the general public concern over these products.
- Lack of Control/ Fear of the Unknown: Considerable discussion took place on this topic, and included concerns over the irreversibility of introducing genetically engineered materials into the environment, the lack of control over such a release, and the potential disruption and impact on the ecosystem.

- Structural Changes in Agriculture: This concern is centered around the structural changes that will likely occur as a result of increased agricultural engineering, and include the evolution of monopolies and consolidation of the food supply through vertical integration, the impact on small to medium sized farming operations, the political implications resulting from a consolidation, and the potential for "new use" agriculture to compromise our ability to feed the world.

Conversely, those features most likely to be viewed as positive by the public include;

- Increased Food Quality: Genetic engineering will provide products with increased nutritional value, longer shelf life, and more cosmetic appeal.
- Improved Food Safety: Advances in technology will decrease the use of pesticides through development of disease resistant strains and improve food safety through longer shelf life and the inclusion of contaminant indicators in food packaging.
- Maintain a Low Cost Food Supply: Higher yielding, disease resistant crops, more efficient processing methods, and longer shelf lives will help to maintain a low cost food supply.

The use or intended use of alternative products raises ethical issues that need to be addressed before acceptance can be achieved. Our responsibilities in addressing the ethical issues in biotechnology have dramatically increased due to the speed of change and the potential for much greater consequences than in other areas. Among the ethical issues to be considered are:

- Playing God: Considerable concern was expressed regarding the issue of the sacredness of life and the right of scientists to alter traditional life forms. Should we cross natural barriers in creating or modifying life forms? Greater concern was expressed over altering animal life forms than plants, but each present significant ethical issues.
- Ownership of Life Forms: Who owns the genes and germplasm resulting from biotech research? Are these life forms patentable?
- Technology for Technology's Sake: Should we use a technology simply because we can? Should we not return to traditional methods? We have a greater understanding of traditional methods, why not use that understanding to achieve our objectives rather than pursuing a controversial technology area?

- Research Funding: Given the controversial nature of this area, should public monies be used to fund research in biotechnology?
- New Product Decisions: Who really decides which products are developed and introduced in the market? Are negative social and ethical issues overlooked in the rush to profits? Should the economic/profit motive be the sole driver, or should social issues factor much more strongly in the equation?

Can regulation provide some level of control over the types of products developed and introduced, and will it help to ease some of the ethical concerns? The public has the right to know the origin of the products they are consuming, whether they are genetically-engineered or contain other additives and by-products. Consumers are asking questions such as "Why are genetically-engineered products not labeled?" and "Does the industry have something to hide?" Legislation requiring explanatory labeling of genetically-engineered food products will have a positive impact on how the industry is viewed. Public opinion plays a key role in acceptance or rejection of nontraditional food products, and can help determine the ultimate outcome through public interest groups, the media, and other public forums.

Economic and Structural Issues

The introduction of new products and processes in the food industry will pose economic and structural issues to all facets of the industry. Discussions on this topic centered around the impacts of new products on agriculture and food processing, how farmers can best prepare for changes, and what partnerships can be developed to take advantage of opportunities. Economic and structural issues are difficult to address, however, due to the broad underlying issues of markets, societal developments, and politics. In addition, dealing with a multinational/global industry on a national level presents tremendous difficulties.

Potential impacts of new products on agriculture being adopted by the marketplace include:

- Horizontal Integration: New classes of agricultural products will provide linkages and involvements with other sectors, including pharmaceutical, energy, and environmental. This will potentially provide additional or modified income sources for producers. This expansion into other industrial areas could also impact negatively our ability to feed populations, should land use be increasingly diverted to nonfood agriculture.
- Vertical integration: Control of food production is likely to lead to consolidation in the industry, which may lead to greater control over

the food supply by the industry and reduced availability in the types and varieties of products. Because of agricultural production exclusively for export use, large multinational farming organizations will likely have less interest in maintaining the overall quality of the land and those communities producing agricultural products, leading to displacement of native populations. Industry, by nature of market forces, tends to take a short-term, profit-oriented approach and does not place significant value on the long-term, public good aspects of their actions.

- More dependable food sources with increased nutritional value and safety will likely result through new agricultural and production processes.
- Resistant Strains: Increased use of biotoxins will likely cause very strong selective pressure for insect and disease resistance, potentially resulting in super strains, and devastating impact on agricultural production.
- "Quick Fix " approaches may distract attention from the broader social and ethical issues, such as more effective distribution of food products to starving populations.

In preparing for this new model of agriculture and food production, the question of what the farming model will look like in 15-20 years needs to be considered. Will it be more highly industrialized with a small number of multinational conglomerates controlling agriculture and producing the bulk of the food supply? Will the small to medium size farming operation survive and thrive? Participants expressed the view that farmers can best prepare for the future by:

- *Staying Informed: Farmers must educate themselves on the technologies available, as well as the political and regulatory climate that exists, and keep up to date on current and anticipated impacts and opportunities.*
- *Get Involved: The voice of the farming community is becoming unified, and will gain effectiveness as the numbers of those involved increases. Participation in the political and developmental process is a necessity.*
- *Consider Alternatives: If vertical integration does occur, farmers must be prepared to seek alternative ways to survive and thrive. These include teaming up to form co-ops or other partnership operations which can compete with large industry, and developing production and distribution alternatives such as community shared agriculture (CSA) and organic farming.*

Biotechnology-based agricultural and food products and processes have the potential for tremendous impact on our society. The question of whether that impact is positive or negative is cause for considerable debate in all sectors. Our society has evolved to the point where we tend to focus on those short-term solutions to the detriment of the potential long-term, global impacts of our actions. We cannot afford to take that approach in the biotechnology sector, as the stakes are too high. A holistic, balanced approach is necessary by all parties involved to assure a proper balance for the future. It is our responsibility to consider the fundamentals first, and continuously maintain a focus on the global issues and impacts in order to assure a successful future. In the area of biotechnology, the definition of a successful future is still open.

Pharmaceutical Industry

Cochairs

Geetha Ghai
Rutgers University, Cook College

Dennis Doyle
Tru-Blu Cooperative

Alan Laskin
Laskin/Lawrence Associates

Rick Walter
Canadian Institute of Biotechnoloy

The purpose of the workshop was to discuss each of the issue areas, determine what impact they may have on the successful development of pharmaceutical applications of agricultural biotechnology, and suggest mechanisms to overcome any identified obstacles. Although most of the participants had little pharmaceutical industry background or direct experience with the applications for human health care through agricultural biotechnology, there were a few participants with direct pharmaceutical industry experience. The participants provided "top of mind" contributions to each of the questions and issues raised.

New Market Opportunities

The workshop explored potential applications of agricultural biotechnology to areas such as health care products, test models, nutraceuticals and functional foods, organs and tissues for transplant, high value proteins and complex molecules.

While it was expected that pharmaceutical applications of agricultural biotechnology may provide biotechnology with an improved reputation, a number of serious barriers to development could be forecast. These include the high cost of development and long lead times, multiple regulatory oversight for the products and processes, poor international regulatory standardization, overly broad patent claims (in the US, Japan and Europe), and the genetic source material being located off-shore.

In order to seize the opportunities, it was generally agreed that the educational system and media must be encouraged to raise the level of awareness in the public. Partnerships will have to be created between the marketing and scientific communities as well as between the numerous groups involved in the research, development, production and sales of the products. The North American public is both poorly informed and skeptical about biotechnology, presenting problems and opportunities. Low levels of trust continue to plague the biotechnology industry; however it was felt that a skeptical public will force the industrial and research communities to improve communications skills and develop more suitable messages.

To create successful market opportunities, it was felt that the initial focus should be on preventative versus curative products. The university research community should consider establishing multidisciplinary teams, as is done in industry, to ensure cross-fertilization of ideas and the development of nontraditional solutions to barriers. Information provision was seen as a priority, with a need for immediate development of education programs to raise the level of public awareness.

Societal Issues, Regulations, and Ethics

Much of the workshop discussion revolved around the public concerns of biotechnology such as the safety of products for the environment and human health, the perceived lack of public/individual control over product development and consumer choice. It was noted that pharmaceutical applications of agricultural biotechnology will likely be seen as beneficial for human health but, at the same time, may be seen as detrimental for animal welfare or environmental impact. The public was seen as supportive of crop and income diversification for the primary producers yet concerned about multinationals controlling both the genetic material and farmers growing the proprietary strains under contract.

While there were only minor regulatory gaps recognized by the workshop, such as the premarket regulation for foods, additives, and supplements in the US, several important regulatory issues were identified. The regulatory scene lacks international harmonization and approval times appear excessive. Since health claims cannot be made on food products (except physiological claims such as dietary fiber), North American regulations are not as supportive of functional foods as other jurisdictions in Europe or Asia. Regulatory change should be considered to meet the increasing consumer demands for access to the functional food category.

Ethical considerations raised related mainly to special-interest groups involved in animal welfare. Animals used as factories for production of complex proteins as well as organ donors for transplant were considered the most likely targets for activist concern. The farmers involved will probably be subjected to increased interest from media as such products become more widely known.

Access to and compensation for countries that own useful genetic resources (mostly the developing nations) was seen to be an issue; however, no specific recommendations were tabled. The Convention on Biological Diversity addresses this issue. The existing agreement between Merck and Costa Rica was considered one possible approach.

Economic and Structural Issues

Pharmaceutical applications of agricultural biotechnology are expected to increase rural job creation, diversify farm incomes with high value crops, and introduce new domestic and export market opportunities. However, in comparison with other applications of agricultural biotechnology, only a small segment of the farm community will be affected by the development of pharmaceutical applications. Relatively few farmers will be required to supply the worldwide market for such pharmaceutical products. Those farmers who are involved will be subject to increased tracking, monitoring and evaluation, and may become closely aligned with, if not employees of, large multinationals.

The development of unusual alliances will result as the traditional lines between agriculture and health care begin to erode. Pharmaceutical, seed, food distribution and food processing companies will work together with health care professionals and farmers. Also, health care funding may begin to be accessed by the agricultural research community. New partnerships will develop for research, education, communications, and marketing. The structure of farms will change with vertically integrated "high tech" small farms. The farming profession will become management intensive with unusual market regulations for farm products such as vaccines and nutraceuticals. The main challenge will arise from quality control product issues.

Information provision was seen as a fundamental requirement to ensure the eventual success of agricultural biotechnology applications. The entire biotechnology community must become communicators by developing concise messages in lay terms, delivering these messages at every opportunity and, where possible, through credible organizations. Mass and electronic media, local government representatives and the educational establishment should be utilized to get the message out. The development of a continuous educational system with credits, similar to the one existing for physicians, should be established for pharmaceutical farmers. Additionally, a group called "Brokers for Communication" should evolve to distribute information not only to the public, but also to the pharmaceutical farmers, educators and others. And, finally, the information materials must include details on the intent of the application or product, how it will benefit the user, the risks associated with its use and the mechanisms in place to manage the risks.

The group's recommendations include:

- *Develop educational instruments to improve the level of awareness in the public, the media, the farming community, and the pharmaceutical, food processing, and retailing industries.*
- *Information materials must describe the intent, benefits, risks, and risk management mechanisms.*
- *Regulatory change should be considered to meet increasing consumer demands for access to the functional food category.*
- *NABC member institutions should promote workshops and other related activities for dialogue between the agri-food and health care communities, academics, and the public.*
- *Participation of industry representatives at NABC meetings should be encouraged.*
- *NABC member institutions should encourage graduate students and post docs to attend NABC meetings by covering their costs.*

PART III

KEYNOTE ADDRESSES

Biotechnology: Catalyst for Change in Agriculture

KENNETH A. BARTON
Agracetus, Inc.
Middleton, Wis.

The modern agricultural industry has been highly effective in the implementation of productive new technologies that provide more food, shelter, and industrial materials for an expanding world population. Agricultural output needed to produce an abundant food supply for nations where economies and distribution systems are adequate is currently achieved on approximately six million square miles of cultivated land, the same land use as in 1950, despite a doubling in world population throughout this period. Such expanding productivity, on constant or even diminishing agricultural land, is an essential component of the sustainable agricultural system we all hope to develop. It can only be achieved through continued scientific discovery to improve germplasm and production practices throughout the world. There is no greater opportunity today to accomplish this than through the application of biotechnology.

Increased global information exchange, international trade, and rapid innovation in many areas of technology — especially biotechnology — are now contributing to an evolution in the world agricultural industry, through which substantial enhancement in our industry will occur much faster than ever before. We will see more rapid improvement and global distribution of germplasm and production practices, and we must prepare to deal with entirely novel applications for agricultural land use. More so than any previous agricultural technology infusion, biotechnology will impact all elements of the industry, all associated industries, and all of interfacing society. The increased speed and the extreme breadth of current change make this a particularly unsettling time. Now, as never before, there is need for industry, academia, and consumer interests to cooperate in the identification and management of issues generated by this new technology. We must deal with new laws and regulations, new products and industry practices, and entirely new agricultural uses that will emerge as the potential of biotechnology is realized.

Two circumstances are important in considering the magnitude of the situation we now face. First is the exceptional power of biotechnology to enable rapid and precise manipulation of plant and animal genetics in ways that were never previously feasible. Second, the application of biotechnology to other industries beyond agriculture ensures that a massive level of research and development expenditure will continue to fuel improvement in biotechnical skills for the foreseeable future. Available resources for continued research and technology development will be well above the historically low levels of funding that have been targeted solely toward agriculture, and advances in other industries will significantly contribute toward the rate of progress in our own industry.

In this presentation I will discuss the breadth of the enabling technology, its management through the intellectual property system, and the effects we should anticipate as our industry adjusts to a more rapid pace in new product development.

The Technology

Biotechnology enables the precise alteration of metabolic processes of living organisms to achieve novel outcomes. Advances in this field have relied upon broad innovation in our understanding and manipulation of living materials, which is achieved through the sciences of molecular and cellular biology. Molecular biology is the study and manipulation of DNA, which is the common "blueprint" of all living organisms. Such strong similarities exist in the character of DNA from all life forms that a common technology base is broadly applicable toward gene discovery, gene mapping and tracking, and even the genetic engineering of diverse microbial, plant, and animal systems. Cell biology, also an integral aspect of biotechnology, is essentially the study of structure and function of living cells. Like molecular biology, many principles of cell biology can be generalized across species.

The necessary laboratory equipment and intellectual skills needed for research in molecular and cellular biology have evolved dramatically over the past two decades, and essential capabilities are already in place in most of the world's major academic and industrial centers worldwide, including not only those which focus solely on agriculture, but other industries as well. Because of the common interest in the application of molecular and cellular biology to the health care, chemical, and other high-value industries, research and development advances toward agricultural objectives will be furthered by discoveries from outside our field. For example, molecular breeding technology, now widely used to facilitate new crop development, continues to benefit greatly from human genome mapping efforts. In addition, the refinement of gene mapping technology will contribute to our increased understanding of the existing mechanisms by which plants and animals adapt to, and deal with, their environments, leading to new opportunities to engineer crop improvements in years ahead. However, it is the transfer of genetic information from one organism to another, even across species boundaries, that represents the most powerful new opportunity for the improvement of plants and animals.

Many schemes have been developed for gene transfer to live organisms and much of the technology is applicable toward any living cell, regardless of species. While plant sciences have often followed the advancement of other industries due to greater financial resources being directed toward fields such as health care, the engineering of plants has progressed considerably faster than comparable efforts in animals or humans because of practical and social considerations. Gene delivery to all major plant species has already been achieved, and the rate of continuing progress in refining crop gene delivery technology is dramatic. With such powerful enabling technology in hand, our industry is now turning to issues of new product development.

Evolution of the Agricultural Industry

We are currently in the midst of an evolution in agriculture, provoked by the maturation of biotechnology from the stage of concept development to application. It is practical to view this evolution in three phases, each spanning approximately a decade. The process began in earnest with a period of concept evaluation and enabling technology development during the 1980s. This was followed by a transition from technology development to early product development, which is now underway. The evolution will continue, following the turn of the next century, with more extensive product development, which will lead to new agricultural applications and dramatic industry expansion.

The 1980s was an extraordinary period of technology enablement during which biotechnical methods were generated for gene identification, gene cloning and characterization, and for the delivery of genes to plants and animals. Hundreds of millions of new dollars were infused into agricultural research — much of it from the private sector — without any immediate financial return through sale of products. It was indeed fortunate for our progress in this emerging field that private sector funding was available, since the early technology development was very costly and occurred as federal research support began to decline. With many of the new technologies now on line, the industry is increasing its focus on product-oriented research, while continuing to refine essential aspects of enabling technology for the future. Although marketable products were not generated during the 1980s, a significant outcome from that period of discovery was intellectual property. The past decade was not unlike the land rush of the past century, and many claims were staked — large and small — that will shape the development of the industry through the ongoing transition.

The agricultural industry in the 1990s will be dominated by a transition from biotechnology development toward product and market development. We are now in the midst of this change, with the first of the new products reaching the marketplace in launches of unprecedented scope. Insect resistant cotton and corn, herbicide resistant crops, and other early products have all been released to a very receptive marketplace. Continued product success will ultimately depend on the level of consistent value delivered to the consumer, but all current indications show that new and faster product development will be a major benefit to our industry. Years of research and product testing have gone into each of the products now reaching the market, and the first revenues from actual product sales are being welcomed by industry investors. However, the development of those new products has further revealed peripheral industry issues that result from the use of biotechnology. The creative revision of market paradigms to enable sufficient value capture from novel products must now be undertaken.

Following the turn of the century, the third decade of this evolution will be characterized by more complex product development and significant commercial expansion. We will watch value shift away from classical agricultural chemical inputs toward more versatile crop and animal genetics. Industry consolidation will continue through alliances and acquisitions, and the trend toward vertical integration will escalate to enable more effective product management and value capture. Many of the small companies that pursued technology development in the past will have insufficient resources to move products to market in a timely way, and many will close or be acquired by larger firms. Companies of greater size and resources will adapt to the changing environment, new production and marketing strategies will be established, and entirely new business areas will be opened to exploit agricultural technology. Among the key elements shaping these industry adaptations to the new technology will be ownership and control of new technologies and resulting products. While various commercial practices will contribute to such control, there will be a strong influence from the patent system.

Intellectual Property in Agriculture

The generation of biotechnology skill has been very costly, and the application of the new technology will entail higher costs in shorter research and development cycles than historically practiced in agriculture. Government and foundation support to finance this technology over the past decade has been very limited. Increasing budget pressures in the federal government will make basic and discovery research dollars increasingly scarce. The situation is ironic: time frames in the enhancement of agricultural productivity are long, which is a detraction to industry, but benefits are seen across society. Government expenditures toward the enhancement of agricultural productivity have also historically provided an exceptionally strong societal return. However, we are facing a period where the continued development of agricultural productivity will fall increasingly on private industry. For effective implementation of biotechnology, product value must be more reliably captured to reimburse even early stage researchers, or the product development pipeline will be diminished. As has been demonstrated in all other technology-dependent industries, patent law provides a mechanism to capture value when products are delivered to consumers. That value is then distributed to contributors along the development path.

A "patent" is a legal provision that provides an exclusive right to inventors for a limited period of time to "make, use, or sell" their discovery in return for releasing the knowledge of their invention to the general public. The patent system has served many other industries well in stimulating research, encouraging product development, and enabling the controlled distribution of products to consumers. Patents are neither new or untested in agriculture, yet their increased application to agricultural products must receive a much greater emphasis in years ahead. Acceptance of a stronger intellectual property system in agriculture has been slow.

Plant "certificate" protection was first offered with the Plant Patent Act (PPA) of 1930, when breeders were given the opportunity to protect asexually propagated crops developed through their breeding programs. Certificate protection was further extended in 1970 with the Plant Variety Protection Act (PVPA), which covered sexually propagated crops. However, an intriguing exclusion of a small class of crops — known as the "soup vegetables," which included okra, carrots, celery, tomatoes, peppers, and cucumbers — provides a useful lesson for our current transition. The soup vegetable exclusion from the PVPA resulted from a concerted lobbying effort by an industry sector that feared proprietary protection would inhibit research and lead to increased vegetable costs. In practice over the following decade, however, the protected crops were the subject of increased research and development, with resulting yield and quality enhancement without unacceptable price increases. In 1980, an amendment to the PVPA was passed in which certificate protection was extended to cover the previously excluded soup vegetables. The lesson learned through a decade of practical experience was one that had already been learned in other industries: effective proprietary protection serves both the agricultural industry and consumers by enabling equitable value distribution for product innovations.

Proprietary protection for crops and animals was further extended when it was determined that utility patent law, which is distinct from the PVPA, can be applied to living organisms. Utility patents address products or processes that are novel, non-obvious, have a defined utility, and are clearly described to the general public, rather than maintained as trade secrets. The extension of utility patent law to engineered plants and animals came about through two landmark court rulings. In *Chakrabarty v. Diamond* (1980) it was determined that specific claims covering engineered oil-degrading bacteria were allowable under utility patent law, because significant human intervention was required to generate the product. This case first clarified that newly created living materials were to be considered patentable products under utility patent law. The relevance to agriculture was further extended in a second case, *ex parte Hibberd* (1985), where it was determined that corn varieties with enhanced amino acid profiles were patentable under utility patent law in addition to what PVPA protection otherwise afforded. The effect of those two rulings was a dramatic escalation of patent filings addressing all aspects of biotechnology. The applications cover many now routine processes such as the use of DNA markers to streamline breeding programs, technologies to make and transfer genes, and the resulting engineered plants, seeds, and transgenic animal products from biotechnology.

Many early biotechnology patents have been issued, but still more remain under prosecution. Concerns have been raised by various parties over the number of new patents and the nature of claims that have emerged, based on fears that agricultural biotechnology patents might impede research and delay future product development. The current concerns are not unlike those which led to the soup vegetable exclusion from the PVPA of 1970. Once again, we can expect to find those fears to be unfounded. Research is proceeding in both academia and industry at a rapid pace, and there has been no effort from industry to extract value from research licenses to academia for the new technology. Indeed, many of the dominant patents in biotechnology are controlled by academic institutions, and license revenues are contributing support for academic research programs. Industry is effectively adjusting to the requirement for patent licenses, and launches of valuable new products have been initiated without inappropriate deterrence from licensing or litigation. The global management of patents is far more unsettled, and it will be many years before consistency emerges, particularly in nations that have not historically protected intellectual property.

The experience from other technology arenas is being followed in biotechnology — early patents addressed broad enabling concepts; later inventions are considerably more limited in scope. This is due to the requirement for novelty in inventions, and as technologies mature it becomes more difficult to achieve substantial advances that previously have not been disclosed to the public. The outcome of this maturation process is that broad patents influence industry development in the early years, while the more limited patents sustain long-term product advancement of the industry. What is unusual in agricultural biotechnology is that many of the most powerful patent applications, filed early in the last decade, still remain under prosecution due to lengthy delays imposed by prosecution backlogs and patent interference proceedings. These dominating patents will have strong influence on the shaping of our industry over the next two decades.

Capturing Value from Biotechnology

Genetic engineering of plants and animals is now well underway, with the essential technology components in hand to develop real products. Gene design and delivery technologies continue to be refined, useful genes for agronomic and production traits are being unveiled in increasing numbers, and product development is now the primary goal of our industry. However, the emergence of the first new products presents a series of complex issues to the industry regarding marketing and value capture.

The agricultural industry has historically been comprised of distinct segments including seed providers, growers, processors, distributors, and consumer outlets. We have generally relied on the passage of commodity materials from the farm, through processors, and on to consumers in distinct steps that capture increasing incremental value as the refined agricultural materials approach the end-user. A modern trend toward larger farming operations and vertical integration, combining processing and distribution, has been driven to date by economies of scale and other efficiencies provided by channeled flow of materials through a controlled development pathway. However, as biotechnology adds increasing value directly to germplasm, it will catalyze a series of changes in industry value-capture paradigms, which will lead to a more comprehensive transition toward identity preservation of crops from seed to consumer.

For agriculture to be successful, products and services must have both a tangible value and a mechanism to capture and return that value to product and service providers. It is common to exchange value at the farm gate through sale of germplasm and chemicals to growers at the processor level, where growers receive compensation for their agricultural produce in return for deliverables to the processor, and at the level of end-users, where processors and distributors receive monetary value from consumers. Biotechnology is now promoting the shift of a substantial component of crop value directly into the germplasm, but this comes at a substantial development expense. Herbicide resistance, pest and pathogen resistance, and various other traits of value to growers, processors, or consumers are already entering the marketplace in the form of new germplasm. These first product examples each involve genes encoding relatively simple, single-gene traits. The traits now being delivered are conceptually consistent with traits which have current market value. However, new mechanisms must be devised to allow payment to be equitably captured and diverted from growers, processors, distributors, and consumers to compensate developers of the novel germplasm.

Herbicide-resistant crops should enable growers to use more effective, ecologically sound and economical herbicides than are currently at their disposal. However, for such products to be effective in the marketplace, the grower must receive a tangible advantage through enhanced productivity and decreased chemical expenditures. It is then reasonable to expect that a portion of the financial savings seen by the grower would be shared with the provider of the novel germplasm. This issue takes on still greater significance when insect-resistant crops are considered. Herbicide resistance shifts farm input expenditures from one chemical product to another, but insect resistance very substantially reduces overall expenditures for purchased chemicals and thereby generates much higher potential savings to the grower. The sharing of this savings with germplasm developers can be accomplished through either payment of a premium on seed sales, or as a direct licensing payment to the developer of the novel trait. However, such equitable value sharing is inconsistent with past industry practices that permit growers to save their own seed for planting in subsequent years, a practice that would naturally limit compensation delivered to those who initially created the novel germplasm. There are now legal restrictions against the reuse of saved seed where the materials are covered by utility patent claims, but industry acceptance must still be achieved. If such compensation cannot be reliably returned to germplasm developers, new cost-saving products will less likely become available to growers in the years ahead.

Faster Product Development

Biotechnology will contribute to the streamlining of product development cycles, or "cycle time," in an industry where advancement has been generally slow and methodical. Plant and animal traits, which in the past were objectives of classical breeding programs, are already being generated in considerably shorter time frames through the use of gene mapping and tracking. Where formerly it may have been necessary to visually observe phenotypic traits in populations of progeny following genetic crosses, it is becoming routine to employ linked DNA markers to screen large numbers of progeny at an early stage of development — even in seed prior to planting — thereby limiting grow out to only the desired progeny. Analysis of the content of donor genome, relative to the recipient, and selection of the most advanced progeny, enables a more rapid introgression in far fewer breeding cycles. Thus, the rate of progress in development of new germplasm has already accelerated, and commercial products will become outdated more rapidly as replacements reach the market in shorter time frames. The shortened product-life will necessitate more effective value-capture mechanisms than in the past.

The physical transfer of genes encoding new traits provides a very powerful tool to further shorten the product development cycle. For example, in programs at Agracetus we have refined "gene gun" DNA delivery technology to achieve delivery of genes directly into seed of commercial crop varieties. This enables the direct germination of transgenic plants without the need for tissue culture, thereby avoiding a process that both delays initial plant development and limits gene transfer to only those varieties which can be managed in culture. Our first transgenic plants, in elite commercial varieties, reach maturity within five months of project initiation and do not require time-consuming back crossing. Because genes encoding new and desirable traits can readily be moved across species boundaries, exceptionally diverse crop characteristics can now be generated in a single growing season.

There are many potential advantages of such rapid product development, but industry adaptation necessary to successfully manage these new capabilities will be complex. This is exemplified in a second example, in which Agracetus scientists have undertaken gene transfer programs to genetically engineer a stronger cotton fiber, a trait that has historically been subject to premium pricing and has been a long-term breeding target. The advantages of increased fiber strength are found at various levels of the fiber industry, from processing steps to consumer satisfaction. Stronger fiber is able to undergo more rapid processing without breakage as the thread, yarn, and cloth are mechanically constructed, and the faster product through-put enabled by stronger fiber converts to immediate savings on expensive capital equipment. In later processing steps, where the chemical and mechanical processes used to "finish" fabrics weaken fiber, a stronger fiber helps retain the durability characteristics that are valued by consumers.

Through classical breeding, the average strength of cotton fiber across the industry has been increasing at a relatively constant rate of 1.5 percent per year, with a cumulative strength increase of 16 percent achieved from 1980 to 1991. However, we are now able to use genetic engineering to dramatically increase the strength of cotton fiber with the addition of genes from other organisms. In our recent research project, the strength of the major upland cotton variety was increased by more than 60 percent with a single transferred gene. This represents a strength enhancement equivalent to 30 years of classical plant breeding in a fraction of that time, and the fiber strength achieved now exceeds the current premium system for fiber strength. Identity preservation of such specialty fibers or an altered premium structure will therefore be needed if exceptional value is to be captured. The industry has effectively adjusted to incremental improvements in many characteristics such as fiber strength and length, but must now consider the economic implications of more dramatic quality changes in shorter time frames. In the fiber industry, and in many other areas where crops are utilized, this will entail a more frequent evaluation of processing technology to efficiently take advantage of new inputs. The capital infrastructure for utilization of agriculture will need to become more flexible in order to accept radically new and varied materials enabled by biotechnology.

NEW PRODUCTS, NEW MARKETS

The ability to bring in new genes from different organisms further enables development of novel agricultural products, including products that will extend existing markets and those that will lead to entirely new business opportunities for agriculture. A second cotton fiber concept from Agracetus offers one example of a product that extends an existing agricultural product line. Our intent in this program is to improve, beyond the capacity of existing cotton germplasm, the dye-ability, chemical reactivity, absorbency, and structural dynamics of fiber (such as shrinkage and wrinkle resistance). Already developed through this program are fibers that contain polyester in the fiber core, and we have found that some of the more desirable characteristics of petroleum-based synthetic fibers are imparted to the "natural" cotton fiber. Similar programs to extend the current limits of materials derived from other crops are underway in many other laboratories where oil, starch, and protein alterations are being advanced beyond the capacity of existing germplasm. The revision of plant products in this manner, by bringing in genetic information from one organism to another, thus represents a broad opportunity to expand current plant product markets. However, such new products will require significant adjustment in processing practices and market structure to enable value capture from traits not routinely monitored by the industry today. Sale of transgenic seed to growers at a premium price will not be sufficient, even as proprietary protection becomes accepted, because the grower has no means to pass on costs downstream to processors. The outcome will be an additional push toward vertical integration, which will facilitate the channeling of identity-preserved products from the field to dedicated processors, and then on to the consumer.

The use of gene transfer between organisms also enables development of entirely new agricultural uses, such as the Plant Bioreactor Program of Agracetus. In this project, crops are used as production vessels to economically produce large quantities of new biological materials. It has long been recognized that plants provide the world's most economical supply of proteins and complex secondary metabolites, but historically the diversity and concentrations of these materials were limited to what was found in nature or developed by man through classical plant breeding. Gene transfer now enables the engineering of crops to produce more of the valuable chemicals that are already found in some species, and also new materials currently derived from other sources. Targets of Plant Bioreactor Production include many high value biological products destined for industrial, food, feed, and even pharmaceutical application. Genetically engineered crops for each of those markets are currently in development in a number of laboratories worldwide, and early product candidates are already in field trials. Because the crops will require special handling in the field, dedicated processing, and delivery to specific end-users, vertical integration will clearly be a favored trend.

SUMMARY

Biotechnology represents a powerful new tool for plant and animal breeding, and the application of this technology will lead to new products and new uses for agriculture. This technology enables very rapid product development. The increased speed of accomplishing genetic improvements also condenses product development expense into much shorter time frames. These issues will necessitate changes in industry practice to ensure that appropriate value can be captured for the increased contribution of genetics as a component of final product value. Historically, high quality genetics have been critical to agricultural progress but have not fared well in capturing a significant portion of final product value. Paper packaging now captures greater value than the grain in a box of breakfast cereal. New paradigms for value capture will therefore evolve to more equitably distribute value from consumers to those participants earlier in the development path. The key industry changes to achieve this will be clarification and enforcement of a strong system for intellectual property protection, and vertical integration of the industry to coordinate value exchange from germplasm provider to consumer. While the most exciting aspect of biotechnology application to agriculture will be novel products and new uses for crops, even commodity agriculture will be affected by broad changes in industry management of crop value.

Novel Crops and Other Transgenics: How Green Are They?

REBECCA GOLDBURG
Environmental Defense Fund
New York, N.Y.

As biotechnology develops, we could see the advent of biological pollution, which could be more dangerous than nuclear or chemical pollution.[1]

Monsanto's agricultural biotechnology provides improvements in quality and yield of crops, and benefits the environment where crops are grown.[2]

Agricultural and environmental applications of modern biotechnology have spurred considerable controversy about their environmental risks and potential to alter the environmental impacts of other technologies and practices, such as pesticide use. The range of biotechnology products under development is expanding rapidly, and thus the potential for controversy over the environmental effects of biotechnology products is also increasing.

The purpose of this paper is to examine environmental issues associated with novel genetically engineered organisms being developed for agricultural, pharmaceutical, and environmental applications. First, I will consider the purposes of these novel biotechnology products and whether environmentalists view these products as environmentally beneficial. Second, I will consider the environmental risks of these novel biotechnology products. Third, I will examine one novel regulatory approach to managing environmental impacts of one category of biotechnology products.

[1]Martin Khor. 1994. Why we need a Biosafety Protocol urgently. Third World Resurgence 48:20.

[2]Monsanto press release. Monsanto receives final regulatory approval for commercialization of insect-protected cotton. October 31, 1995.

ARE NOVEL BIOTECHNOLOGY PRODUCTS GOOD FOR THE ENVIRONMENT?

Biotechnology products frequently are touted as environmentally beneficial. Even in the face of such claims, however, it is unclear to many environmentalists whether biotechnology products will be good for the environment. In part, this is simply because of the increasingly varied nature of biotechnology products, which range from industrial chemicals to transgenic animals. It is unreasonable to expect that as a class, biotechnology products will be "good" or "bad" for the environment. But even individual biotechnology products or types of biotechnology products face skepticism from the environmental community, for at least two reasons.

The first reason is that the environmental community has grown wary of environmental claims in general. As public support for environmental protection has grown, corporate environmental claims for products have become more common. Unfortunately, these claims are not always legitimate. Plastic supermarket bags, for example, sometimes are touted as "ecological," even though the bags are not made from recycled materials and customers frequently dispose of the bags after one use. The practice of telling consumers that particular products are good for the environment, even when their environmental benefits are dubious, has become so common that the environmental community has given this practice a special name: greenwashing. As greenwashing has become common, so has legitimate skepticism of environmental claims — including claims for biotechnology products.

Second, skepticism about environmental claims for novel biotechnology products is also based on past experience with hype concerning the potential of biotechnology. In the past, for example, biotechnology's promoters have promised that fertilizers will become unnecessary as crops are engineered to fix their own nitrogen, and that pesticides will become obsolete as crops are engineered to resist insects and other pests. Those unrealistic claims have spurred general caution about environmental promises for biotechnology products.

As a specific illustration of the environmental community's response to biotechnology products, consider bioremediation, a technology that has been promoted heavily as one way in which biotechnology will aid environmental protection. Most environmentalists support the concept of bioremediation — harnessing natural processes to degrade hazardous chemicals. Nevertheless, the environmental community has not rushed to embrace bioremediation as the solution to problems with hazardous wastes.

In part, this response is based on the potential of environmental "snake oil" to be disguised as a legitimate bioremediation product. Following the oil spill from the tanker Mega Borg off the coast of Texas in 1990, for example, one Texas company promoted heavily the success of its bioremediation product — bacteria that supposedly "eat" oil on the surface of the ocean. According to a report published by the Texas General Land Office, these bacteria largely dissipated an oil slick from the Mega Borg in just seven hours, with portions of the slick breaking up in just 30 minutes. The speed of this degradation is difficult to believe, especially given that the experimental design had no replication of treatment and control areas, and because the oil in the treatment area simply may have been dispersed by wind and water. The relatively low cost of such bioremediation techniques makes them attractive to government agencies and companies that must remediate waste, but also signals caution to many environmentalists. Reliance on inexpensive but unproved bioremediation products could cause considerable environmental harm if the result is that more efficacious cleanup methods are not used.

Not just skepticism, however, has caused bioremediation's relatively low profile in the environmental community. Many environmentalists are now focused on pollution prevention rather than remediation of wastes. Changing industrial processes to minimize the amount of waste produced is regarded as the best way to end the problems caused by chemical wastes. Over the long-term, biotechnologists may do more for the environment by developing novel enzymes and other tools that allow the redesign of industrial processes, than by developing bioremediation methods.

DO NOVEL BIOTECHNOLOGY PRODUCTS POSE NOVEL RISKS?

As of May 1996, government records indicated that in the United States there had been about 3,500 field tests of genetically engineered plants, 50-100 field-tests of genetically engineered microorganisms, and two field-tests of genetically engineered fish. These numbers continue to grow, involving an ever greater diversity of genetically engineered organisms. I will examine the environmental risks associated with genetically engineered products, as these risks apply to species or taxa that only recently have been genetically engineered.

Crop plants: Within the predominant group of engineered organisms — plants — the range of species being field-tested until recently has been relatively narrow. As of March 1993, about 85 percent of field-tests of genetically engineered crops were of six species: corn, cotton, tomatoes, potatoes, soybeans, and tobacco. However, the diversity of crops being genetically engineered has increased substantially: transgenic varieties of more than 40 different crop plants have now been field-tested in the United States. These include fruits and vegetables such as cranberries, papayas, raspberries, and radicchio; ornamentals and turf plants such as chrysanthemums, gladioli, petunias, and creeping bentgrass; and trees such as poplars, spruce, and sweetgum. Moreover, the range of traits being introduced to crop plants is also increasing. Along with insect, disease, and herbicide resistance — the traits most commonly introduced to crops — a number of crops are now being engineered to produce pharmaceuticals, polymers, and industrial enzymes, and to alter oil, starch, and protein contents.

The prominent ecological risk associated with crop plants is that they will transfer, via pollination, their acquired genes to related wild or weedy plants, or to other cultivated non-transgenic varieties of the same crop. How will these gene transfer risks apply to the wider range of crop varieties now under development?

The answer to this question is not straightforward. Gene transfer to wild or weedy-related plants may pose both lower and higher risks. The primary concern about such gene transfer is that the transferred genes may encode a trait that confers a selective advantage. Populations of wild or weedy plants that acquire the trait may increase to the extent that they become an ecological or agricultural nuisance. Acquired traits for insect, disease, and herbicide resistance — the traits that have long dominated genetic engineering of crop plants — realistically could confer a selective advantage to a wild plant. However, it is difficult to envision many other traits now being engineered into crops, such as production of pharmaceuticals, polymers, and industrial enzymes, and altered oil, starch, and protein content, as conferring a selective advantage to a wild plant. Therefore, transfer of such traits to wild or weedy plants generally should pose low risks.

On the other hand, the wider diversity of plant species now being engineered could increase the likelihood of gene transfer to wild or weedy plants. Concerns about gene transfer by most of our major crops — the traditional focus of genetic engineering efforts — are minimal in the United States. Most of these crops originated in other areas of the world and do not have wild relatives in the United States. Genetic engineers, however, now are focusing increasing attention on plants such as forest trees and ornamentals that have wild populations in the United States. Many of these plants readily can transfer their acquired traits to wild relatives, and thus have the potential to pose significant gene transfer risks.

Gene transfer from transgenic varieties to cultivated non-transgenic varieties of the same crop on average may pose higher risks than in the past. Gene transfer to cultivated, non-transgenic varieties to date has not been a focus of great concern, because most introduced genes are intended for use in food crops and are intended to be safe for consumers. For example, consumers likely would not be exposed to significant health hazards if a small percentage of a corn crop acquired a herbicide-resistance gene from a transgenic corn growing nearby. (One important exception to this generalization would be the transfer of genes encoding substances that are safe for most consumers, but allergenic for a minority of consumers, which would pose a serious risk for the allergic minority.)

In contrast, genes encoding pharmaceuticals or industrial chemicals are not intended to be used in foods, and expression products of such genes could pose health hazards to consumers. Transfer of such genes from transgenic to non-transgenic varieties could contaminate food derived from the non-transgenic varieties, posing health risks to consumers.

Fish: Sharp declines in wild fish stocks and new technologies have created strong economic incentives for aquaculture (fish farming). In the United States, aquaculture production more than doubled between 1984 and 1993, and aquaculture is now the fastest growing segment of agriculture. Worldwide, aquaculture is growing by about 10 percent each year. Farmed shrimp, for example, now are more than 30 percent of the world shrimp supply.

Interest in engineering aquatic organisms for aquaculture is growing rapidly, potentially creating new environmental risks. One United States company, AF Protein of Newton, Mass., has approached the United States Food and Drug Administration for approval to commercialize Atlantic salmon engineered to produce a growth hormone from chinook salmon. According to AF Protein, their transgenic salmon fry grow 400-600 percent faster than their non-engineered kin.

Such transgenic fish arguably pose some of the greatest risks of any type of transgenic organisms. Aquaculture facilities are notoriously leaky. Data from waters off Scotland's Faroe Islands, for example, indicate that as many as 48 percent of the salmon caught by commercial fishermen have escaped from salmon farms. In general, fish escaped from aquaculture facilities are reasonably likely to survive and breed with natural fish populations, because most fish have not been debilitated significantly by domestication. Therefore, transgenic fish readily may transfer their engineered traits to wild fish populations, where these traits might spread via natural selection. Having acquired an advantageous trait, fish then could affect populations of other aquatic organisms — for example, by competing for food. These sorts of ecological effects are of concern particularly because considerable experience with introductions of "exotic" fish from other geographic areas suggests that aquatic ecosystems are highly vulnerable to ecological disruptions.

Invertebrates: Recently, a number of invertebrate species have been engineered for the first time. In 1996, the USDA received the first applications for field tests of transgenic mites and nematodes. In aquaculture, shrimp and mollusks were first reported as transformed. Ecological risks posed by introductions of transgenic invertebrates have only begun to be examined. Those risks especially merit scrutiny because many invertebrates are highly mobile at certain life stages and have extremely high reproductive rates, and because transgenic invertebrates readily may interbreed with wild conspecifics.

NOVEL REGULATORY APPROACHES TO MANAGING ENVIRONMENTAL IMPACTS OF AGRICULTURAL BIOTECHNOLOGY

Innovative new regulatory approaches may help our society better manage the environmental risks and controversies associated with an expanding universe of biotechnology products. One potentially helpful regulatory tool is conditional registration of pesticides. For some products, conditional registrations may strengthen environmental protection while providing significant advantages for regulators and manufacturers.

In order to market a pesticide in the United States, the substance must be registered by the U.S. Environmental Protection Agency (EPA) under the Federal Insecticide, Fungicide, and Rodenticide Act (FIFRA). In general, a company submits considerable data to the EPA concerning the human health and environmental impacts of their product. Based on these data, the EPA decides whether to register the pesticide. If the EPA chooses to conditionally register a pesticide, the agency continues to require the submission of data over a certain time period after the pesticide is commercialized.

FIFRA was originally amended to allow conditional registrations in order to eliminate a double standard that arose as a result of increased data requirements under the statute. Conditional registrations enabled the EPA to allow manufacturers to start selling a pesticide while working to fill data gaps, as long as the EPA determined that the conditional registration would not increase significantly the risk of any unreasonable adverse effect on the environment.

The EPA's use of conditional registrations took an unprecedented turn in March 1994, when the agency granted a landmark registration to two agrochemical companies — Monsanto Company and Zeneca AG — for the herbicide acetochlor. The EPA conditionally registered acetochlor on the basis of arguments by the herbicide's manufacturers that the availability of acetochlor would greatly reduce overall herbicide use on midwestern corn crops. Amid considerable controversy, the EPA weighed the manufacturer's arguments for environmental benefits against other evidence that acetochlor is carcinogenic and may leach into ground water. Taking an unprecedented step, the EPA registered acetochlor, but established a number of conditions for continuing the registration beyond an initial five year period. Monsanto and Zeneca must document promised decreases in herbicide use and must fund programs to monitor ground water and surface water for acetochlor contamination. If the companies do not meet these conditions, the EPA will cancel the pesticide's registration.

The EPA subsequently issued several conditional registrations involving genetically engineered crops. In May 1995, the agency conditionally registered bromoxynil for use on cotton that was genetically engineered to tolerate this herbicide. This registration was highly controversial because bromoxynil is absorbed through the skin and is a reproductive toxin, thus posing health risks to farm workers. However, bromoxynil's manufacturer argued that use of bromoxynil-tolerant cotton would lead to a decrease in overall use of more hazardous herbicides on cotton. Similar to acetochlor, the EPA conditioned bromoxynil's registration on the requirement that the manufacture substantiate its claims of a net environmental benefit.

Later in 1995 and in 1996, the EPA conditioned registrations of corn and cotton genetically engineered to express insecticidal Bt toxins from bacteria. Preparations of naturally occurring Bt bacteria already are used as a safe insecticide on relatively limited acreage by both conventional and organic farmers. Many scientists are concerned that genetically engineered Bt crops will be planted widely, leading insect pests rapidly to evolve resistance to Bt toxins, and rendering Bt useless in both genetically engineered crops and as a natural insecticide. The EPA conditionally registered Bt corn and Bt cotton to require that the crops' manufacturers establish programs to prevent or slow the evolution of insect pests' resistant to Bt.

Recent conditional registrations are distinct from previous conditional registrations in the scope and kind of data they require. They are not appropriate for all products, because they involve some extra time and effort on the part of agency and company staff. However, they can be a "win-win" regulatory solution in certain situations. Conditional registrations have at least four compelling advantages:

1. *Conditional registrations build into EPA's regulatory program for pesticides powerful new incentives for companies to evaluate carefully and honestly the environmental implications of their products. If a company cannot document within a specific time period the performance it promises for a pesticide — e.g. that Bt-resistance will not evolve — the company's registration may be canceled.*
2. *The fact that companies must document their claims, which otherwise are easy to exaggerate, should give regulators considerable new confidence in the information companies provide and should boost shaky public confidence in regulatory decisions concerning pesticides.*
3. *Manufacturers should gain from the EPA's formal consideration of a company 's environmental performance predictions that now are not generally considered as part of the registration process.*
4. *Manufacturers gain orderly decisions allowing the commercialization of controversial products that otherwise might be tied up in lengthy regulatory debates.*

In short, conditional registrations are not a panacea. They have the potential, however, to enable the EPA to establish strong incentives for companies to develop pesticides that provide net environmental benefits and to practice environmental stewardship in the management of pesticide products.

Conclusion

The expanding universe of biotechnology products will broaden the range of environmental risks and controversies associated with biotechnology products. However, the diversity of new products means that while some will become the focus of new concern and debate, others may arouse little interest from environmentalists. Innovative new regulatory approaches are one way in which our society may manage better those products that spur environmental controversies.

PART IV

PLENARY LECTURES

Genetic Engineering of Flavor and Shelf Life in Fruits and Vegetables

David A. Evans
DNA Plant Technology Corp.
Oakland, Calif.

Twenty years ago, consumers only ate fruit that was "in season." Fruit was not shipped long distances and supply was dependent on local production areas. Demand for year-round, inexpensive produce has led to consolidation of production where temperatures are warm, sunshine is plentiful, and labor is cheap. In most cases, produce is now shipped long distances before being consumed. Tomatoes grown in Mexico are trucked 3,500 miles, bananas grown in Ecuador travel 6,000 miles on a boat, bell peppers from Holland greenhouses are flown 5,000 miles to market. Unfortunately, fruit was not designed to be shipped long distances. In order to make it to market, fruit is picked before it is ripe (and before it has any taste). Despite harvesting immature fruits, post-harvest losses of fruits and vegetables still exceed 25 percent of crop production and the fruit that makes it to market has often been described as tasteless.

Strategies for Flavor and Shelf-Life Improvement

Flavor can be addressed by increasing intensity (concentration) of an important chemical component of flavor. This can be accomplished by increasing production of a specific compound (e.g., sucrose), by shifting the balance of compounds (glucose versus sucrose), or by blocking production of undesirable compounds (such as starch). Genes have been isolated to explore increased sweetness. These include sucrose phosphate synthase (increased sucrose), invertase (interconversion of sugars), and ADPG-pyrophosphorylase (sugar to starch conversion).

Alternatively, flavor can be improved by delaying harvest until the fruit is ripe. Greater firmness or reduction in ethylene production can permit fruit to be retained on the vine until ripe, when it has achieved full flavor potential. Increased firmness through regulation to decrease fruit softening enzymes has met with limited success to date. Control of ethylene content appears to be more promising. Ethylene controls ripening and rotting in many fruits and vegetables. By reducing ethylene concentration in fruit, it is possible to arrest development of the fruit and permit the fruit to remain on the vine until ripe.

Case Study: Ethylene Regulated Tomatoes

Technology

Tomato fruits produce ethylene in high concentrations during ripening. There is a burst of ethylene production, called the climacteric, that occurs when the tomato fruit begins to turn from green to red (Grierson and Covey, 1988). The increased concentration of ethylene in tomato fruit is associated with color development, fruit softening, increased respiration, and sugar accumulation. Ethylene is a simple chemical. Its biosynthesis is understood well and the genes that are required for biosynthesis have been identified. In addition, genes have been identified that degrade intermediates in ethylene biosynthesis.

Ethylene formation is the result of a three-step biosynthetic pathway (Imaseki, 1991): Methionine is converted to S-adenosyl-L-methionine (SAM) by the enzyme methionine adenosyltransferase. SAM is converted to 1-amino-cyclopropane-1-carboxylic acid (ACC) by the enzyme ACC synthase (ACC-S). ACC is converted to ethylene by the enzyme ACC oxidase. Genes encoding the enzymes involved in biosynthesis (ACC synthase and ACC oxidase) have been identified, isolated, and cloned. It is possible to use techniques to block expression of those genes, thereby blocking or reducing biosynthesis of ethylene. Alternatively, enzymes have been identified in microorganisms or viruses that metabolize the intermediates in ethylene biosynthesis. These include ACC deaminase, an enzyme capable of degrading ACC, and SAM hydrolase, an enzyme capable of degrading SAM.

The process for developing genetically engineered plants with reduced concentrations of ethylene in the fruit is complicated and requires several technologies that have been developed over the past twenty years. These technologies include gene introduction into plants, selection of transformed cells, regeneration of plants from genetically transformed cells, and gene expression.

Commercial Opportunity

The principle commercial benefit of ethylene-regulated tomatoes is long shelf-life. Tomatoes blocked for ACC-synthase can survive in laboratory conditions for 90 days. At DNA Plant Technology Corp., we completed commercial tests on a hybrid tomato variety that was blocked for ACC-S using Transwitch™ technology. In commercial tests, these ethylene-regulated tomatoes have a 40 day shelf-life. This includes harvest, five to seven days ethylene treatment to complete color development, five to seven days to ship to market, five to seven days storage in the local market at a repacker, seven days in the supermarket, and 10 days with the consumer prior to consumption. This contrasts with the maximum of 20 days for current commercial tomato varieties.

The current commercial system works as follows (How, 1991):

1. Growers harvest a tomato field two or three times during a season. All tomatoes (unless red ripe) are harvested.
2. Tomatoes are brought from the field to a packing shed where they are sorted by size and color. Tomatoes with some red color are usually sold at a discount price in the local market because they are already beginning to soften and will not survive trucking over long distances. The 90+ percent green tomatoes are placed in 25 pound boxes and treated for one to three days with 100-150 ppm ethylene, then trucked to a repacker at a local market.
3. Upon receipt, the repacker sorts tomatoes for color. Tomatoes that developed color in transit are shipped to supermarkets (or food service outlets) immediately. The remaining tomatoes receive an additional two to seven days of ethylene treatment and then are shipped to market.

Extended shelf life provides a distinct advantage throughout the chain of commercialization.

Grower: The ACC-S regulated tomato reaches full size and flavor potential, begins to degrade chlorophyll, but does not turn red in the field. At this stage, the tomato can remain in the field for several weeks without rotting or softening. This attribute affords the grower two benefits. First, all the tomatoes are harvested, i.e., no over ripe tomatoes are left in the field. This increases field yield per acre. Second, harvest costs per pound are lower because fewer total harvests are required.

Packer: Over-ripe tomatoes are sold at a discount in local markets (Florida, California, and Mexico). Because all the ACC-S regulated tomatoes that are harvested can be shipped to remote markets, revenue per pound is increased. In addition, tomatoes left on the vine generally increase in size. Large tomatoes command an average premium price of four to six cents per pound.

Repacker: When tomatoes arrive at a repacker they are sorted based on stage of ripeness. Losses at the repacker range from six to ten percent of tomatoes shipped from production areas. These are discarded tomatoes that have rotted in route or are showing signs of impending deterioration. Ethylene-regulated tomatoes are superior, with decreased losses.

Retailer: Because current tomatoes rot in two to three days at retail, orders and shipments are made daily by most large supermarkets. Loss due to rotting can reach 20 percent of product at the retail store level. Both issues are addressed by a long shelf-life tomato. Moreover, better tasting tomatoes can be sold at a premium price, resulting in increased revenue and profit per square foot in the produce section of the supermarket.

Consumers: Because ethylene-regulated tomatoes are left on the vine until they reach full maturity, the consumer purchases a better tasting tomato. Moreover, there should be a significant reduction in the number of rotten tomatoes in consumer homes.

The commercial potential of a better tasting, longer shelf-life tomato that is sold at a premium price can be quantified as follows:

per 100 lbs. harvested	standard	ethylene controlled
Revenue	$20.00	$60.00
Costs	$16.00	$38.00
Margin	$ 4.00	$22.00
Margin %	20%	37%

INTELLECTUAL PROPERTY

The technology that is required to make an ethylene-regulated tomato is the subject of several patents and represents a mine field that must be successfully navigated prior to introducing a product to market. The present intellectual property situation is as follows:

Germplasm: Nearly all commercial acreage of fresh market tomatoes is grown using proprietary seed developed by a seed company. The open pollinated, publicly available varieties (e.g., Rutgers, Floradade) cannot compete with current hybrid varieties in yield and disease resistance, hence genetically engineered open pollinated varieties are not commercially viable. As proprietary seed is controlled by seed companies, relationships must be negotiated with seed companies to obtain access to parental seed for genetic engineering. The largest tomato seed companies are Seminis, Ferry Morse, and Sun Seeds. Other companies with breeding capability have developed proprietary tomato germplasm using breeding or cellular genetics (Morrison and Evans, 1996). In addition to hybrids and the protection inherent in segregation of hybrid seeds, most companies also seek Plant Variety Protection Certificates on one or both of the hybrid's parents. Those layers of protection, along with seed production capability, assure seed company involvement in commercialization of genetically engineered tomatoes.

Gene Introduction: There are two methods available for gene introduction in tomatoes. *Agrobacterium tumefaciens*-mediated gene transfer is the preferred method. *Agrobacterium* plasmids are modified to incorporate desired promoters and genes and the natural infection behavior of the bacterium is used to deliver the DNA into plant cells. Broad claims have not been issued on this technology. However, several components of this technology, such as the use of binary vectors, are the subject of patents. A secondary gene introduction method is ballistics. Ballistics uses some method to shoot DNA into plant cells. Variations on the type of microprojectile and the acceleration device have been optimized during experimentation for different crop plants and likely would need to be optimized for tomatoes. DuPont Corp. has claims in a U.S. patent that are directed to the ballistics technology. Other patents are issued or pending on various modifications of the ballistics approach for specific plant species.

Plant regeneration: Regeneration using standard tissue culture methods is routine in tomato. The method using leaf pieces of tomatoes as explant material has been used in several laboratories since the early 1980's. No broad patents cover the commonly used techniques for tomato regeneration.

Promoters: A wide range of promoters have been identified that will express DNA in plants. Some of these, including 35S, derived from cauliflower mosaic virus, are the subject of issued or pending patents (Fraley, et al., 1994). Much effort has been directed to isolation of tissue-specific promoters. Fruit-specific promoters in tomato would be ideal to ensure that ethylene regulation is directed to the tomato fruit and to ensure that ethylene production (involved in disease resistance) is not blocked in leaf and other tissue.

Selectable markers: Markers are foreign genes that are inserted into target plant cells at the same time as a segment of DNA that can modify ethylene. Because ethylene regulation cannot be selected in a test tube, the selectable marker permits transformed cells to proliferate while non-transformed cells die. Resistance to kanamycin (conferred by the nptII gene) is a preferred selectable marker that has received both EPA and FDA approval for use in tomatoes and other crop plants. Kanamycin resistance is the subject of an issued patent in the United States and a pending patent in Europe (Rogers and Fraley, 1991).

Genes: Segments of DNA can be patented in the United States. For ethylene, patents are issued or pending on each gene that has been identified in ethylene biosynthesis or degradation. For example, the USDA has a pending patent on the ACC-S gene that we have used. The gene isolation work was completed at the USDA in Albany, California (Oeller, et al., 1991). The USDA has licensed the gene to multiple companies within different fields to ensure broad commercialization of the government funded technology. Other genes have been isolated including ACC-oxidase (Hamilton, et al. 1990), ACC-deaminase (Klee, 1994), and SAM hydrolase (Good, et al., 1993).

Gene expression: In order to block ethylene biosynthesis using ACC-synthase or ACC-oxidase, the endogenous genes must be regulated to decrease their expression. Two methods have been used in tomatoes: Transwitch® and antisense. Transwitch® is a unique phenomenon in which some plants resulting from transformation with a homologous gene are suppressed for the target gene (Jorgensen, 1995). Antisense achieves a similar result, i.e., gene suppression, by inserting a gene sequence that is complementary to the target gene (Shewmaker, et al., 1992). Each technique is the subject of issued patents in the United States and Europe.

Commercialization Implications

Despite the demonstrated value of ethylene-regulated tomatoes, no one is presently in the market with such a product. This product has extremely high barriers to entry due to the intellectual property situation. The companies closest to market are DNA Plant Technology, Monsanto/Calgene, and Zeneca Plant Sciences. It is likely that each party will need licenses to intellectual property from other companies in order to successfully commercialize an ethylene-regulated tomato. While the current patent situation presents an additional commercialization hurdle to the three companies, it also tends to eliminate the prospect of future new entrants developing and marketing an ethylene-regulated tomato.

Research and Planning Implications

Several general lessons have been learned from attempts to commercialize ethylene-regulated tomatoes that apply to all commercial projects using plant genetic engineering.

1. Many patents are still pending. Work initiated today using technology developed by others may never be commercialized in the absence of licenses. Many of the basic agricultural biotechnology patents have not been resolved, with broad patents still possible or likely for gene introduction and gene regulation technology.
2. In commercial or academic genetic engineering research, genes need to be inserted into the best existing germplasm. This affords an appropriate benchmark to value the technology and eases adoption of new varieties. Seed company and technology-provider alliances involving universities and research laboratories are more likely in the future.
3. The earliest product developed is not always the first to be commercialized. The complex business system of fresh market tomatoes and the complex patent situation have delayed commercialization. If a new genetically engineered variety requires modification of growing or handling practices, adoption is slow.

4. In a complex business system a novel trait may have multiple benefits. The ethylene-regulated tomato provides benefits to growers, packers, repackers, retailers, and consumers while originally targeted to increase shelf-life.

Summary

The initial hurdles for commercialization of products of plant genetic engineering included technical, regulatory, and consumer acceptance. As the first products are now ready for market introduction, it is obvious that an additional major hurdle exists: intellectual property. Patents have been granted that cover a number of basic biotechnology methods and the use of specific genes for genetic engineering. For some technologies and genes, patents already have been issued and represent barriers to commercialization. For other technologies, the patents are still pending and represent an uncertainty. Indications from the first products being commercialized in cotton, corn, and soybean suggest that patents are being aggressively enforced and are being used to establish competitive advantage in the marketplace. Clearly, if broad patents are upheld by the patent office and the courts, alliances and cross licenses will be necessary for commercialization.

The modification of consumer-preferences traits could have implications for several, perhaps unintended, steps in the chain of commercialization. For example, ethylene regulation in tomato will primarily result in a tomato with longer shelf-life. However, the tomato will not be red when it reaches full maturity, can stay in the field for a longer period of time prior to harvest, requires post-harvest ethylene treatment, and results in delivery of a better tasting tomato to consumers. These features of the ethylene-regulated tomato require modification of harvest practices and modification of handling and packing procedures. Moreover, production trials and operation tests may suggest modification of pesticide practices, staking, and handling in the field; sorting, ethylene treatment, and shipping from the field; as well as packaging and pricing for the retailer. Ultimately, successful commercialization of new consumer-preference traits in fruits and vegetables requires more than simple insertion of a gene into a plant.

REFERENCES

Fraley, R.T., R.B. Horsch, and S.G. Rogers.1994. Chimeric genes for transforming plant cells using viral promoters. U.S. Patent #5,352,605

Good, X., J.A. Kellogg, W. Wagoner, D. Langhoff, W. Matsumura,, and R.K. Bestwick. 1994. Reduced ethylene synthesis by transgenic tomatoes expressing S-adenosyl methionine hydrolase. *Plant Molecular Biology*. 26:781-790

Grierson, D. and S.N. Covey. 1988. *Plant Molecular Biology*. pp. 114-115. Blackie: Glasgow and London

Hamilton, A.J., G.W. Lycett, and D. Grierson. 1990. Antisense gene that inhibits synthesis of the hormone ethylene in transgenic plants. *Nature* 346:284-287

How, R. B. 1991. *Marketing Fresh Fruits and Vegetables*. pp. 104-113. New York: Van Nostrand, Reinhold

Imaseki, H. 1991. The biochemistry of ethylene biosynthesis. In: *The Plant Hormone Ethylene*, A.K. Mattoo and J.C. Suttle, eds. Boca Raton, Fla.: CRC Press.

Jorgensen, R.A. 1995. Cosuppression, flower color patterns, and metastable gene expression states. *Science* 268:686-691

Klee, H.J. 1993. Ripening physiology of fruit from transgenic tomato (*Lycopersicon esculentum*) plants with reduced ethylene synthesis. *Plant Physiology*. 102:911-916

Morrison, R.A. and D.A. Evans. 1996. High pigment, reduced blossom end scar size, disease resistant tomato varieties. U.S. Patent #5,489,745

Oeller, P.W., L. Min-Wong, L.P. Taylor, D.A. Pike, and A. Theologis. 1991. Reversible inhibition of tomato fruit senescence by antisense RNA. *Science* 254:437-439

Rogers, S.G. and R.T. Fraley. 1991. Chimeric genes suitable for expression in plant cells. U.S. Patent #5,034,322

Shewmaker, C.K., J.C. Kridl, W.R. Hiatt, and V. Knauf. 1992. Anti-sense regulation of gene expression in plant cells. U.S. Patent #5,107,065

Plant Cell Culture Technology

M. DIANNE DEFURIA
Bristol-Myers Squibb Company
Princeton, N.J.

The rationale for utilizing plant cell culture for production in the pharmaceutical industry can be exemplified by the realities faced in TAXOL® development. Such a case study provides the justification for Bristol-Myers Squibb's (BMS) investment of well over $10 million to develop this technology to the scale of commercial production. BMS believes that other products will follow for plant cell culture and that other new production technologies will be developed as the need and opportunity are identified for high value products.

Why Plant Cell Culture for Pharmaceuticals?

- Historical success and future promise of natural products as therapeutics
- Increasing constraints on bioprospecting and wild biomass collection
- Emerging technologies increase utility as research tool and feasibility for commercial use
- Identification of high-value products with supply problems

Why plant cell culture in the pharmaceutical industry? You are familiar with the historical successes and what we hope is the future promise of natural products with therapeutic applications. The pharmaceutical industry realizes that there are increasing constraints on bioprospecting, as well as the collection of wild biomass for production of natural product-based therapeutics. There are many emerging technologies, some of which David Evans described earlier, that have inspired people to use plant cell cultures as a research tool for drug discovery, whether from direct isolation, biotransformation, or directed fermentations. BMS is the first to look at cell culture for commercial production of a large, complex therapeutic molecule because of its high value and supply problem. The compound is paclitaxel; the active ingredient in the product, TAXOL®. This product alone motivated a large pharmaceutical company with massive investments in all kinds of production facilities to turn to plant cell culture — a recent scientific development with little practical application.

A more in-depth look at the four reasons for using cell culture in the pharmaceutical industry reveals that the full potential of natural products in medicine has not been realized.

Success and Promise of Natural Products

- 75% of world population relies on plants for treating illness/disease
- 25% of U.S. pharmaceutical market from plant-derived compounds, including state-of-the-art drugs, e.g. *TAXOL®*
- Only 2% of the >250,000 plant species have been extensively evaluated as therapeutics
- Unparalleled diversity of complex, novel molecular structures

Seventy-five percent of the world's population depends almost entirely on plants to treat illness and disease. Historically, 25 percent, and now somewhat less, of the U.S. pharmaceutical market has been based on plant-derived compounds. There are the older remedies such as digitalis, but also state-of-the-art drugs like TAXOL®. Very few of the two or three hundred species of plants have been specifically evaluated for their therapeutic benefit, and the compounds that have been derived are very novel, diverse, and complex. There are, undoubtedly, many more plant-derived compounds of medicinal value to be identified. However, one important question arises: "*Can the continued traditional screening of plant materials economically compete with new combinatorial chemistry approaches?*" The answer is not yet in, but many are betting on a combination of the two technologies.

Constraints on Bioprospecting

- Geo-political impediments to access
- Difficulty of reliable resupply
- Unrealistic expectations of many source countries
- Development vs. preservation of biodiversity, disappearance of rainforest, extinction of many species

The constraints on bioprospecting are many, (see Figure 3). BMS and many other companies have screened natural products for many years. We have all experienced the increasing geopolitical impediments to accessing new biomass for screening. This is because of the difficulties of prospecting in countries with underdeveloped scientific infrastructure, less stable governments, or other constraints. The need for resupply after an interesting compound is identified may follow years after acquisition of the original plant material. Scientific or government personnel changes may have occurred, and access for resupply is sometimes denied or impossible to accomplish. Many source countries have unrealistic expectations. Plants and their genetic material have been viewed as "green gold." Many countries believe bioprospecting within their borders offers a rare opportunity for economic development in a context of limited possibilities. This view is unrealistic based on the potential discovery of a single pharmaceutical product after 3-5 years of screening and another 10 years of high risk development by the pharmaceutical company. In addition, there is the need to balance the problem of disappearing rain forests and extinction of many plant species with the need for development in source countries. As seen in Figure 4, many technologies, from emerging genetic-engineering possibilities to evolutionary improvements in standard research techniques, have been used to increase the utility of cell culture at any scale.

Emerging Technologies

- Transgenic plants proteins (Abs, insulin)
  - *Agrobacterium spp* vectors
  - "Gene Gun" technology
- Improvements in analytical chemistry, robotics, and "micro research"
- Improvements in bioreactor design for enhanced mass transfer

The last underlying reason for cell culture in the pharmaceutical industry is identification of a high value product with limited options for facile commercial production. TAXOL® clearly fits into this category.

High Value Products/Supply Problem

- Skikonin—naphthoquinone for skin ailments and as a dye in cosmetic and silk industry
- Paclitaxel (*TAXOL®*)
  —diterpenoid for cancer therapy

TAXOL®

Pre-CRADA TAXOL® Development

1964	Anticancer activity of Pacific yew bark extract noted
1971	Isolation and purification reported
1979	Unique mechanism of action described
1983	Clinical trials initiated
1989	Activity in refractory ovarian cancer reported

TAXOL® was identified over 30 years ago by the National Cancer Institute (NCI). They found the compound to be a drug with exciting potential for cancer therapy, but sufficient quantities for development hindered its progress for decades.

In the early 1960's, an extract of Pacific yew bark was found to contain very low concentrations of the active material. In addition, supply of the bark was extremely limited. It wasn't until 1971, prompted by this limited supply, that Dr. Monroe Wall isolated and determined the structure of TAXOL®. In 1979, Susan Horovitz identified that paclitaxel killed cancer cells in a unique way, by binding to tubulin. Paclitaxel's novel structure and unique mechanism of action made it a high priority for clinical trials by the NCI.

The progress of the clinical trials was extremely slow because only 0.5 kilogram was produced over two years. Although trials were started in 1983, it took until 1989 to demonstrate clinical activity in solid tumors. The first success was with refractory ovarian cancer. The NCI wanted to proceed, but needed a partner to further develop the drug. A competition was held to determine who in the private sector could best develop the drug. Bristol-Myers Squibb won the competition and then established cooperative agreements with multiple government agencies, not a trivial task, to continue development.

The USDA's Forest Service and the Interior Department's Bureau of Land Management assisted in an extensive bark collection program in the Pacific Northwest — mostly in large areas of Washington, Oregon, and Idaho. The government agencies developed plans to harvest the bark in a responsible way, to transfer the bark to BMS, and to minimize the impact on the environment of this huge collection to assure that the yew tree was not threatened. Bristol-Myers Squibb paid for all those activities, bought the bark, and managed the bark collection contract with Hauser Chemical Company. No one had experience with such a large collection effort, but it was well worth the effort since the drug was made available to the NCI for vastly expanded clinical trials within a few months.

TAXOL®

Cooperative Agreements

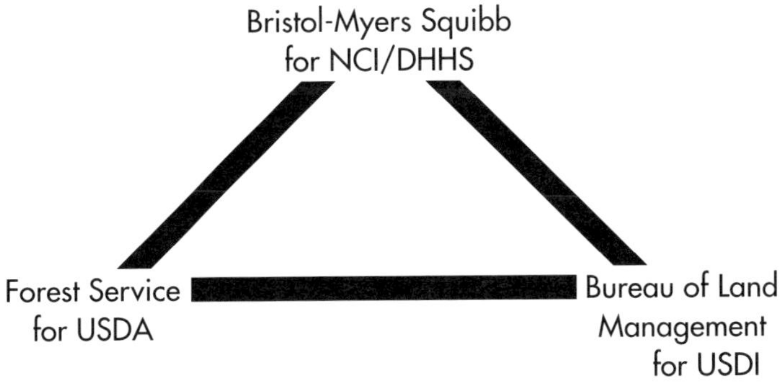

- Government Agencies—to develop annual plans, oversee bark transfer, and ecosystem research/conservation studies
- BSM—pay all government expenses, buy bark, and administer bark collection contract

The Pacific yew tree, *Taxus brevifolia*, is a small, scrubby-looking, extremely slow-growing tree. Very little was known about it — whether it existed and how many there were. The tree was considered a nuisance shrub by timber companies harvesting high-value timber like Douglas Fir. Fortunately, from spring to fall the bark of the Pacific yew was easily peeled by hand from the trunk and major limbs of the tree. Up to 1,000 people were in the woods harvesting bark from 1991- 1993. The Forest Service issued a permit for each collection bag and policed the effort closely. There was quite a bureaucracy associated with this collection and the prevention of theft. The bark was taken to collection centers where it was ground to a workable size, further milled, either dried in the sun or in low temperature tumblers, and then packed in large wooden boxes and trucked to Colorado to be chemically extracted. This process was a low-tech but high brute-force effort. It was an inefficient process, with uncontrollable quality of the raw materials, but at the time it was the only possible production process and it worked to the benefit of U.S. public health.

An alternative source was needed as quickly as possible, so BMS immediately established the multifaceted research program outlined in the following figure.

TAXOL®

Alternative Sourcing Research

- Yew plantations
  - — R&D to determine best cultivar and conditions
  - — Cultivation of biomass for commercial supply
- Renewable biomass (clippings)
  - — Precursor isolation/semi-synthesis
  - — Direct isolation of paclitaxel
- Plant cell culture
- Total synthesis ?
- Other ??

Together with the Weyerhauser Paper Products Company, several yew tree plantations were established that served two purposes — research for determination of the best cultivar for future plantations, and development of the best methods to cultivate the huge amounts of biomass that would be necessary for commercial supply. Prior to reaching these objectives, BMS had 12 million yew trees planted as a commercial source of biomass. The research effort with Weyerhauser used tens of thousands of clippings collected from ornamental nurseries all over the United States. They were hand-planted and cultivated for months in huge greenhouses built specifically for this purpose so that all growing conditions could be completely controlled. The plants were transferred to nurseries to acclimate them to the outdoors and then grown in commercial-scale nurseries. They were grown for three to five years before harvest. The initial idea was to harvest the entire plant and process it to extract either the core of the paclitaxel molecule or the intact product. But, it turned out that the important precursor 10 deacetyl baccatin III (10 DAB) was much more readily and economically available from *Taxus baccata,* which grows in Europe and Asia, than from the North American Taxus species. Therefore, commercial supplies of 10 DAB were produced in Europe for use by BMS.

Years earlier the company had licensed a process to attach the precursor 10 DAB to the active side chain, and semisynthesis proved to be a commercially viable option. The renewable biomass, originally gathered in the wild in Europe and Asia, but now cultivated in Italy, is extracted and the purified 10 DAB is shipped to Ireland where the side chain addition occurs. The purified bulk paclitaxel is then shipped to Puerto Rico for formulation and the product, TAXOL®, is shipped to our distribution centers worldwide.

TAXOL®

TAXOL® Alternative Supply

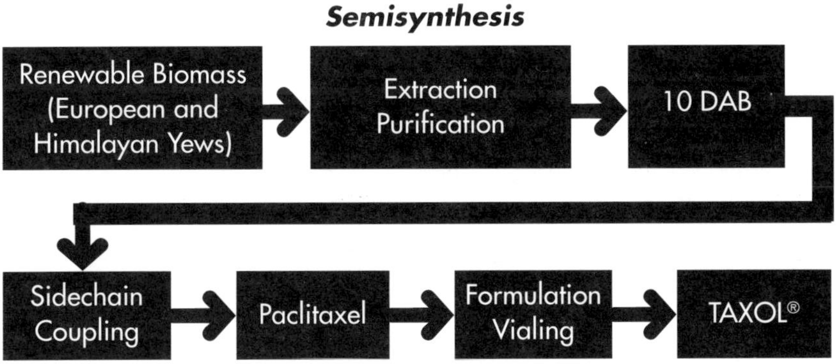

While the above process is much more efficient and controllable than bark extraction, the vagaries of weather can still have a significant impact on total production and quality of the raw materials.

The plant cell culture option was brought to the attention of BMS by a small company, Phyton Catalytic. This company had been interested in developing plant cell culture specifically for the production of paclitaxel for several years. BMS agreed to sponsor their research and licensed the technology as a possible method of production.

Might the BMS collaboration with Phyton on plant culture provide an even better process? We believe the answer will be yes. Although a great deal of research has been accomplished, much development remains. Tens of thousands of yew tree explants were grown on solid nutrient media until calluses were developed. Small flask suspension cultures were used to select and determine optional growth parameters for high producing cell lines. Further research was done in the bench top scale bioreactor where it was determined quickly that paclitaxel produced in cell culture had fewer impurities than that isolated from bark and needle extracts. In addition, cell lines could be selected to produce a specific Taxane of interest.

The potential advantages of cell culture production of paclitaxel emerged quickly and can be generalized to other fermentations as well.

Advantages of Plant Cell Culture

- Environmentally benign
- Faster growth compared to plants
- Controlled, reliable supply of high quality bulk
- Quick response to variability in demand
- Culture conditions controlled easily
- Simplified downstream processing
- Novel metabolites

Plant cell culture is a more environmentally benign process and produces product faster than cultivation of biomass. It is a controlled, reliable source of high-quality material and production levels can be matched to commercial demand. The cell culture conditions are more easily controlled to assure the fermentation of the product of choice and its high quality. Downstream processing for isolation and purification of the desired product often can be simplified. This may translate into lower costs, but that has not yet been established. Cell culture provides the opportunity for additional novel metabolites to be identified that may have a broader spectrum of activity or an increased potency. These interesting possibilities have been recognized by major pharmaceutical companies other than BMS. For example, Merck and Pfizer are using cell culture for discovery of new or improved medicinal compounds, and new companies, such as Phytera, are being formed specifically for this purpose.

Today, in Arensburgh, Germany, Phyton operates the world's largest dedicated plant cell culture facility. There are cascades of fermentors from 75 to 75,000 liters, and the facility is currently being modified to comply with Good Manufacturing Process regulations required for production of human pharmaceuticals. BMS and Phyton continue to collaborate on scaling up the fermentation of new *Taxus sp.* cell lines and to develop an optimized isolation and recovery process.

Today, TAXOL® is approved for use in the therapy of both breast and ovarian cancer. New, improved treatment schedules have been developed and approved for use. The semisynthetic method for manufacturing the drug is approved and TAXOL® is currently in use in more than 50 countries worldwide. Another major market, Japan, should clear the product for marketing in 1997.

TAXOL® Bristol-Myers Squibb

Regulatory Status

- U.S. FDA approvals
  - 12/92 ovarian cancer
  - 4/94 breast cancer
  - 6/94 improved dosage regimen
  - 10/94 semisynthetic manufacturing process
- Approved in over 50 countries worldwide

BMS expects the need to increase somewhat as the clinical research program defines new uses for the drug. The company is confident, however, in its ability to meet all patent and market demand for TAXOL®. New efficiencies will be implemented in the semisynthetic process, as well as full development and commercialization of plant cell culture within a few years.

TAXOL® is an exciting product, which in addition to offering clinical benefit to hundreds of thousands of patients, has been the sole justification for commercial development of an important technology that was little more than a laboratory curiosity for decades.

The Environmental and Energy Sector and Agricultural Biotechnology

JEFF GAIN
AARC Corporation/USDA
Hardin, Ill.

I have been referred to as the father and even grandfather of modern-day industrial uses of agricultural materials. In January 1985, in Columbia, Mo., Roger Mitchell and I had the first "big" debate/discussion on the industrial uses of agricultural products. This was in the period following the agricultural crisis of the late 1970's and early 1980's. The question was "Can we do something with agricultural products other than eat them or feed them to livestock and then eat them?" We needed additional non-food and non-feed uses for our excess production of agricultural crops. We had to find some new uses for this "stuff."

At a reception in Washington D.C. at about that same time, I ran into Secretary of Agriculture John Block. He asked what I would do if I were secretary and I responded that I would appoint a group to look at agriculture with a new perspective — what can we do with the stuff we grow other than use it for food and feed? We had several million excess acres at that time with which we needed to do something. Secretary Block appointed a task force called the Secretary's Challenge Forum, which led to the New Farm and Forest Products Task Force. That task force met throughout a two and a half year period. I was privileged to be a member of the group. We made a series of recommendations that ended up on Capitol Hill. I was fortunate to work with Cooper Evans, who was on President George Bush's staff for agricultural policy. The staff was supportive, and the 1990 farm bill included legislation for the establishment of the Alternative Agriculture Research and Commercialization Center (AARC) to invest in commercialization of non-food and non-feed uses of agricultural and forestry materials. In the 1996 farm bill, the AARC Center became the AARC Corporation. The AARC has made investments in more than 60, mostly small, entrepreneurial companies that use agricultural and forestry materials to make value-added industrial products. Those products include construction and building materials, oils, lubricants and fuels, paper, landscaping, composting and plant protection materials, environmental remediation products, shipping and packaging materials, household and personal care products, and human oral health products. The list is diverse and represents categories from niche to huge markets like fuels. I will provide more details to document the reality of this industrial products opportunity. Biotechnology is not yet playing a major role in this area, but is expected to do so.

An intense education effort has been necessary to promote this industrial products opportunity for agriculture. This has involved the Congress, governmental agencies, and environmental groups. The promising story of industrial uses, including the whole area of biotechnology, has been told. It involves the participation and opportunity for growers to benefit or profit beyond traditional mass-produced raw material kinds of markets; the full use of our arable land; the replacement of imported materials such as petroleum with domestically produced plant materials; the creation of new jobs in rural areas and associated rural development; and the environmental benefits of using plants rather than petroleum. The AARC's mission and strategic plan addresses this area in greater detail.

Let me put a historical perspective on industrial use of agricultural materials. These efforts were initiated in the 1920s, 1930s, and 1940s by Henry Ford, Thomas Edison, and Billy Hale, who married Mr. Dow's daughter (of Dow Chemical), and Wheeler McMillan, who was the well known publisher of the *Farm Journal*, a prolific writer, and overall interesting guy. McMillan, who was 98-years-old, spoke at a meeting in Washington in the late 1980's. He gave one of the most dynamic presentations that I have ever heard. In 1933, that group representing the Chemergy Council, came to Washington and tried to establish some permanent farm legislation to give an industrial agricultural approach to farm programs. They tried to mandate 10 percent use of ethanol — does that sound familiar? The American Petroleum Institute was formed about that same time. We have gone in the direction of teaching and emphasizing petroleum in our universities and have almost eliminated any emphasis on the use of plant or bio-based carbon. Our national energy policy needs to refocus from almost exclusively coal, petroleum, and gas to a reemphasis on bio-based materials.

The reemphasis has started. We're finding a lot of information on the shelves of laboratories like the USDA Regional Laboratory in Peoria, Ill., in companies, and various other places. Some of it is being developed, but it is not known because of the confidentiality concerns. It is exciting to see that finally we're getting underway. I believe we're beginning to find ways to put rural America and agriculture back to work in a meaningful rural economic development program. I want to give you a few specific examples.

We have formed a group called the North American Industrial Hemp Council, and we have made very serious efforts to keep the "recreational" users out of it. It is a significant example of a rare opportunity for agriculture. Industrial hemp is not used as a source of marijuana. Marijuana comes from a different hemp plant, not the one grown during WWII for industrial use. Industrial hemp produces a fiber that can be used to make paper and clothing. Because of the unfortunate association of marijuana with hemp, there are legal restrictions that prohibit its growth but allow its sale in the United States, and allow its growth, but not its sale, in Canada.

We need to take a serious look at those issues. If New Jersey wants to do something, let me talk to you about generating a bill for your legislature. We have done four of them this year — Colorado and Missouri did not make it, but Vermont and Hawaii passed bills. Essentially those bills permit test plot production to provide material for evaluation of uses, how to handle it, and how to market it — all the questions that need to be answered for a new crop and marketing of its products. The Drug Enforcement Agency (DEA) says that you cannot grow it unless you have a cyclone fence and guard towers and searchlights.

Let me tell you about the opportunity for a crop like hemp. I'm working on behalf of the AARC Board with a former U.S. senator who represents a company that manufactures the interiors of vans and automobiles, like side panels and dashboards. They use a wood fiber from Spain and they don't like it. It's not the quality they want plus it doesn't meet Detroit standards for a "green" car by 2002. Ford Motor Company says we want ours by 1999. This company is the sole supplier of those parts for Ford, General Motors, Chrysler, Honda, and Toyota. They are looking at kenaf, which is a tropical crop; hemp also meets their needs. Paper may be another opportunity for hemp. It's increasingly unpopular to cut down trees to make paper, and we're using more and more paper. The demand is up and rising in the paper industry. International Paper has a member on the Board of Directors of the North American Industrial Hemp Council; they are very concerned about alternative sources of paper. They need a good, renewable, annual, large volume supply of a quality long fiber, which hemp provides. Hemp can be grown anywhere. It's also extremely resistant to disease and insects. It does require high levels of fertilization, but it anchors the soil with its great root system. It reduces the need for pest control and it's one of the most resistant plants in the world. Norman Borllaug, Nobel Peace Prize winner, told me in St. Louis recently that he would do anything he could to help make this plant available to grow industrially because it's a fantastic plant. It's the oldest known commercially grown plant in the world. It has lots of things going for it. Does biotechnology have a role to play in hemp?

In the AARC we are seeing the commercialization of agriculture products that appear to have significant positive environmental impacts, as well as agricultural and economic development impacts. One example is a product called Citrasolve being marketed by a company in Connecticut. Citrasolve is a hand cleanser made from citrus. The product may have other uses, such as control of difficult-to-control fire ants, and a university is examining that possibility.

Maybe this product could be a biodegradable kind of soap that you could spray on tomato, bean, or other plants to control pests. What are the applications of some of these products? We really don't know. What we are finding is that the more you look, the more you find.

We are also finding that marketing, not technology, is the major challenge for bio-based agriculture and forestry products. There are many opportunities and it is difficult to decide where to focus. Marketing is the biggest challenge in conducting due dilligence of projects for possible funding by the AARC. We have lots of good ideas, but it is hard to get them into the market. We usually don't have the marketing intelligence that we need.

In the energy area, liquid fuels is a major market that is currently dominated by gasoline from petroleum, more than 50 percent of which is imported. Ethanol from corn has grown to one to two percent of the gasoline market, where a mixture of ethanol and gasoline is sold as gasohol. In the long-term, a combination of waste paper, straw, residues, and a crop such as switch grass grown for biomass may provide ethanol that is cost competitive with liquid fuels from petroleum. The pollution laws in certain states make it necessary to find ways to use/get-rid-of straw because it cannot be burned and there is too much to plow under. Ethanol production from those kinds of waste streams makes a lot of sense and will help clean up the environment. AARC has made investments for commercialization of ethanol from biomass and biodiesel. Some biobased industrial products need technical advances to be economic, as does ethanol, while others are economic and need production and marketing input.

In the 1980s I thought that economics would be the biggest problem. I don't think so anymore. Lubricants from plants are an environmentally favorable example. Crankcase oil for engines and oil for chain saws, outboard motors, and hydraulic fluid lines are being made from plant oils. Because the plant-based ones are more biodegradable than petroleum based ones, we are going to be required by law, like in Europe, to use them. Some biobased industrial products are going to be driven by environmental requirements and others are going to be driven by the pure economics of the marketplace. We need to project what those opportunities/needs will be so that agriculture and forestry material use can be maximized.

My grandparents and parents composted everything and they had fewer pest problems. We are relearning the benefits of composting. AARC has funded Earthgro to commercialize a compost that has disease suppressive ability and reduces need for chemical fungicide sprays. This product has added economic value and is favorable to the environment.

Of the about 30,000 plants that are known, agriculture has only domesticated and used about 30. Soybeans is a new crop to the U.S. that has become very important, and canola is a genetically modified crop that has become important in Canada. Why don't we have more crop alternatives? There is a lot of opportunity out there for new crops and products from them. The more I look, the more I'm convinced of this opportunity. AARC has made some investments in new crops such as Syrica (milkweed) and kenaf, also for their fiber.

The Defense Department is a driver because of national security and because of its increasing emphasis on the environment. It is essential to make linkages to those interests outside of agriculture. Those linkages are essential to marketing. The size of the United States, in contrast to Japan, will enable us to develop approaches to things here that are not possible for countries like Japan. The United States has the acreage to support biobased liquid fuels while countries like Japan do not. But we must establish linkages with major groups outside of agriculture to meet the challenges.

Let me close with a factual statement about plants. I will use corn as a plant example. You can make anything out of a bushel of corn that you can make out of a barrel of crude petroleum oil. The former is renewable, the latter is not. The processes for corn are different than those for petroleum. Petroleum now dominates energy and chemicals. We need to refocus use of plants such as corn for those products. The plant route will be more favorable to the environment and to rural growth and economic development. Biotechnology has a significant role to play in this redirection.

The Public and Agricultural Biotechnology: Key Questions

CARON CHESS
Rutgers University
New Brunswick, N.J.

When I speak to managers of environmental agencies or chemical companies, they are often looking for a risk communication magic wand. The hope is that I will provide a few risk communication abracadabras that will make their audiences see risk as the managers do. In short, managers are hoping for a risk communication incantation so audiences, which are overreacting to a risk (in the eyes of the experts), will magically calm down. And, they assume I can also provide another incantation to wake up audiences that are overly complacent about other risks. I have been studying communication about toxic chemicals for many years and have yet to find any magic wands. While much of the research on toxins has applicability to agricultural biotechnology, I can safely say that you are even less likely to get magic wands today. Instead, I plan to discuss five questions that I feel have relevance to the communication issues facing agricultural biotechnology: 1. What the role of information? 2. How does support differ for specific products? 3. Who supports agricultural biotechnology? 4. How can trust be increased? and 5. Can our institutions handle the rapidity of technological change? My examples come from research dealing with chemicals and radiation, but I think you will find them applicable to agricultural biotechnology.

WHAT IS THE ROLE OF INFORMATION

I'll begin by challenging your notion about the power of information. I think many of us have in our heads a model that says if you provide people with information you will change their attitudes and, in turn, their behavior. This model (see Figure 1) is a bit simplistic. Yet, this is the model that drives the biotech industry to "educate" the public in hopes that the public will then support field trials of biotech plants, will buy biotech products, and will favor the biotech industry.

FIGURE 1:

Overly Simplistic Model of Human Response

Information ⟶ **Attitude** ⟶ **Behavior**

Let's look at some research on the role of information. The data indicate that people are far more complex than the linear model that suggests an injection of information will transform how people respond to agricultural biotechnology. If people were as simple as the model in Figure 1, no one would smoke, we would all use our seat belts, and I certainly wouldn't jump horses on the weekends. We have all experienced this at some level. If you have tried to "educate" your spouse or your children, you have found that information doesn't necessarily transform the family's behavior. Information is not sufficient to make people rational. A social scientist (Rayner, 1992) has pointed out that the concept of "rational" is subjective: where you sit in society determines where you stand on what you consider rational.

Because I am an academic, I am going to provide you the results of several empirical studies about the relationship between information and behavior. The nuclear industry, concerned about the public's "irrational" concerns about nuclear power, has funded a variety of studies to determine if well-informed people support nuclear issues. Most studies focused on what the people knew about radiation. Questions asked were of the following type: What is the process that generates energy in nuclear power plants? What is the fuel that is used in nuclear fission plants? The researchers then looked at the relationship between knowledge of radiation and pro-nuclear attitudes.

About half of the studies indicated that the people who knew the answers to such questions supported nuclear power (Johnson, 1993). The other studies either found no difference in knowledge between pro- and anti-nuclear supporters or found that the people who knew more were more anti-nuclear. Johnson (1993) also reviewed studies dealing with other issues such as irradiated food and hazardous waste. He found, once again, that some studies indicated that knowledgeable people supported the technology and, in other studies, the relationship between knowledge and support was just the opposite: the more knowledge, the more opposition to the particular technology.

The research on the relationship between support of agricultural biotechnology and knowledge seems to show the same signs of inconsistency. An early survey by the Office of Technology Assessment (OTA) (1987) found approximately the same percentage of support for agricultural biotechnology among those who were college graduates and those with less than a high school education. However, a more recent survey of New Jersey residents (Hallman and Metcalfe, 1993) found that support for genetic engineering was approximately 80 percent among the college educated or those with some college, while less than 60 percent support was found among those with a high school diploma or less than a high school education. And both the OTA survey and the more recent New Jersey one found significantly more support among those who said they had heard a lot about genetic engineering than those who heard little. Conversely, an analysis of attitudes of citizens of different European countries towards biotechnology, found that countries with the highest level of education and information, Denmark, Germany, and the Netherlands, had the least support for biotechnology (Almas and Nygard, 1995). In short, the link between knowledge, attitude, and behavior is unclear, at best.

Am I suggesting that you can forget about providing information? No. But I am suggesting that experts of all types tend to overestimate the transformative power of information.

What Constitutes Useful information?

Let's look at what is considered useful information. I wonder if the studies found tenuous links between knowledge of radiation, attitudes, and behavior, in part, because the radiation experts' notions of important information are a bit skewed. If you need to make informed decisions about nuclear power, how important is your knowledge of the energy process and the fuel? Similarly, the Centers for Disease Control was very upset that a survey they conducted early in the AIDS epidemic suggested that people didn't know AIDS is caused by a virus. But Baruch Fischhoff, one of the country's most eminent scholars concerning risk communication, pointed out that changing people's behavior is not going to depend on whether they know transmission is caused by a virus rather than bacteria.

One of the questions for the biotechnology industry is: What should stakeholders know about agricultural biotechnology to make informed decisions? This is the question you need to consider carefully. In fact, you should conduct research on what people want to know about agricultural biotechnology. I have read many studies about people's perception of biotechnology in general, and agricultural biotechnology in particular. None of them asked people what they wanted to know about agricultural biotechnology. While questions were asked about their attitudes towards products, no one was asked if there was any information that they needed. And yet, there are a variety of materials written for the public about agricultural biotechnology. Did anyone ever test those materials to see if readers cared about, let alone understood, the information?

Should Industry Rethink Labeling?

For the most part, the agricultural biotechnology industry has vehemently opposed labeling of its products. But results of surveys of the public suggest the opposite. Even supporters of agricultural biotechnology feel strongly about the desirability of labeling (e.g., Hallman and Metcalfe, 1993). Arguably, the question of labeling needs to be considered as a way to address concerns about biotechnology. If you want me to know more about the technology, the products might come with a label, like that on my yogurt container that tells me about acidophilus. In fact, there is a large body of research that suggests people see risks as riskier if the risks are unfamiliar (e.g. Slovic, 1987). This explains why people are usually more fearful of chemical plants than automobiles — despite the mortality statistics that indicate chemical plants cause far fewer deaths per year than automobiles. The evidence suggests that familiarity with new technology does not breed contempt, but rather greater familiarity and comfort.

People learn when they have a reason to seek information. For example, we are motivated to learn because we need to make a decision or we are emotionally involved in some way. You can hold all the conferences you want, but you will not get most people to attend because they do not yet care enough about the issue to devote the time. So how will you reach consumers? Yes, marketing is important. But you might want to look at labeling to help create both familiarity and information-seeking behavior. Questions to consider about the role of information: Are you over relying on information as the way to change behavior? What information do people want? Can labeling help?

How Does Support Differ for Specific Products?

The research on agricultural biotechnology indicates that people view different products differently. They don't view biotechnology animal products, for example, the same way they do vegetables developed through biotechnology (Lacy et al, 1991) Research is beginning to look at the specifics. See Table 1.

Table 1. Support for Agricultural Biotechnology Products

Percentage of respondents that strongly or mildly approves of the product

New grass that doesn't need to be mowed so often	78 percent
Fruits and vegetables that are less expensive	73 percent
Better tasting fruits and vegetables	67 percent
Fruits and vegetables that last longer on the supermarket shelf	57 percent
Hormones that enable cows to produce beef with less cholesterol	57 percent
Hormones that allow cows to give more beef	39 percent
Hormones that allow cows to give more milk	40 percent

(Data from Hallman and Metcalfe, 1993)

Opinion varies by the type of product, and the perceived benefits of that product may impact the formation of that opinion. Respondents felt no shortage of beef (at least not sufficient to overcome their objections to genetic manipulation of animals) but did care about whether they could eat hamburgers with less cholesterol. Less expensive vegetables are supported more than vegetables with longer shelf life. Those responses make sense when you think about what matters to consumers.

Other products raise questions about how the risks and benefits might be distributed through the population. Grass that needs less mowing appealed to most people (at least in May when the survey was conducted). From another perspective, New Jersey residents spend a great deal of money on lawn mowing services that employ workers that might otherwise be unemployed. Also, the services are run by people who count on the spring and summer to provide the bulk of their income for the year. The extent of support for various products raises a critical question: Who benefits from new use agriculture and who bears the risk? (Lacy et al, 1991).

WHO SUPPORTS AGRICULTURAL BIOTECHNOLOGY?

I have been exploring the differences between people who support the concept of agricultural biotechnology products and those who don't. The New Jersey data in Table 1 suggests that some people were supporters of products that didn't exist at the time of the survey. What is the difference between those early supporters and those who say "no thank you" to a product that does not yet exist?

Multivariate analysis of the data using discriminate analysis suggests that the issue of morality is one of the strongest discriminators between those who oppose a product and those who support it. In addition, for men, one of the key variables that distinguished those who supported agricultural biotechnology products was their conviction that they had already eaten biotechnology products when that was not yet possible in 1992. These data reinforce the notion of familiarity reducing fear. The men obviously did not know much about the agricultural biotechnology market, but they thought they did. They may have been uninformed, but they were reassured by their own perceptions of reality. Thus, another question to explore: What makes people supporters of agricultural biotechnology? This recurring issue of familiarity raises once again the question of labeling.

HOW CAN TRUST BE INCREASED?

The perception of trust in the agricultural biotechnology industry is important: What makes people feel trusting. One of the founders of the field of risk perception has studied this issue — using nuclear power as the basis (Slovic, 1993). He asked people to respond to statements about a hypothetical nuclear power plant in their community. He gave two versions of the same statement — one phrased in a way to decrease trust and another dealing positively with the same issue. Figure 2 illustrates how these two questions were framed. The statement: "the county medical examiner reports that the health of people living near the plant is better than average" had a minimal impact on trust. But the trust-decreasing statement had a very powerful impact.

Figure 2. Judged Impact of a Trust-Increasing Event and a Similar Trust-Decreasing Event[a]

	Impact on trust						
	Very small 1	2	3	4	5	6	Very powerful 7
Trust-increasing event The county medical examiner reports that the health of people living near the plant is *better* than average	21.5	14.0	10.8	18.3	17.2	16.1	2.2
Trust-decreasing event The county medical examiner reports that the health of people living near the plant is *worse* than average	3.0	8.0	2.0	16.0	21.0	26.0	24.0

[a]Cell entries indicate the percentage of respondents in each impact rating category.

In Table 2, Slovic presents data that show what we always have known: Trust is very easy to lose and very hard to build. Trust-decreasing events have a significant negative impact on participants' trust, and the trust-increasing events minimally increased trust.

Table 2. Differential impact of trust-increasing and trust-decreasing event. Note: Only percentages of Category 7 ratings (very powerful impact) are shown here.

Equally significant is the one statement that increased trust more than minimally: A local board has the power to shut down the plant if it is not performing up to expectations. In essence, the statement conferred power to the community to make the hypothetical plant live up to public expectations.

How should you involve people in making decisions so you increase trust? Peter Day and Laura Meagher developed a community advisory panel in southern New Jersey before the field trial of genetically-engineered eggplants. The involvement of stakeholders in decision-making was successful in increasing trust by incorporating their suggestions.

The key questions are: What power will consumers have? What power will the government have? And to be even more provocative, might you be better off if you said: Go ahead. Regulate us. We know we can do it. Would this type of willingness to yield power inspire greater trust? Essentially, this was the route taken voluntarily by the manufacturers of the Flavr-Savr™ tomato to increase consumer confidence. Research should be done to determine if more such actions will increase trust in the agricultural biotechnology industry.

Rapidity of Technological Change

Last night I was both exhilarated and terrified by information about the rapidity of the development of agricultural biotechnology products. The speaker expressed frustration with the slow rate of government responses to these advances. Yet, consider that we have essentially the same government infrastructure we had 25 years ago, and we have universities operating for the most part as they did years ago. Our decision-making capabilities as a society have not evolved significantly, as the difficulty of developing environmental policy illustrates. In short, our societal institutions are not even beginning to keep pace with our technology. We need to pay more attention to bridging the gaps between technological innovation and institutional capacity. I recommend to you one small step towards bridging the gap: a 1996 report by the National Academy of Sciences — *Understanding Risk: Informing Decisions in a Democratic Society*. You might examine this report for suggestions on how to bridge the social and technological issues confronting agricultural biotechnology and responses to some of the questions I have raised. Dealing with the social issues of agricultural biotechnology deserves at least as much attention as technological issues.

REFERENCES

Almas, R., and B. Nygard. (1995) . *New biotechnologies: Attitudes, social movements, and regulation. A comparative analyses of a European discourse.* Prague: European Congress of Rural Sociology

Hallman, W. K., and J. Metcalfe. (1993). *Public Perceptions of Agricultural Biotechnology: A Survey of New Jersey Residents.* New Brunswick, N.J.: Rutgers University's Ecosystem Policy Research Center

Hallman, W. (1996, January). Public Perceptions of Agricultural Biotechnology. *Resource*, pp. 12-14

Johnson, B. (1993). Advancing of Knowledge's Role in Lay Risk Perception. *Risk: Issues in Health, Safety, and the Environment*, 190-212

Lacy, W. B., L. Busch, and L.R. Lacy. (1991). Public Perceptions of Agricultural Biotechnology. In B. R. Baumgardt and M. A. Martin (Eds.), *Agricultural Technology: Issues and Choices.* West Lafayette, Indiana: Purdue University Agricultural Experiment Station: pp. 129-163

Rayner, S. (1992). Cultural Theory and Risk Analysis. In S. Krimsky & D. Golding (Eds.), *Social Theories of Risk.* Westport: Praeger. pp. 83-116

Slovic, P. (1987). Perception of Risk. *Science* (236), 280-285

Slovic, P. (1993) Perceived Risk, Trust and Democracy. Vol 3. No 6 675-681

staff. (1987). *New Developments in Biotechnology*. Background paper, public perceptions of biotechnology. Washington, D.C.: Office of Technology Assessment (90)

Journey to the Origin: Biological Integrity and Agriculture

MIRIAM THERESE MACGILLIS
Genesis Farm
Blairstown, N.J.

At the outset, it is important for me to describe myself as a layperson in this gathering. My background is in art. I have no formal education in science or agriculture. But I am here with a point of view completely opposed to biotechnology in general and to its place within agriculture in particular.

I say this out of my own experience. I left teaching art in the early 1970's, having been deeply moved by the specter of world hunger as it was looming on the global horizon. After the oil embargo of the same period, and in order to meet its balance of payments, the United States quadrupled the price of its export agricultural commodities. It was a response to the lawlessness of the global market system, whereby unilateral decisions on oil and food commodities wreaked havoc on the desperate economies of the non-industrial world. I remember the vivid realization of learning that the cattle of the feedlots of the United States were daily consuming grain adequate to feed the hungry peoples of the world. That fact determined the direction of my life, and I have been involved with the analysis of hunger and agriculture ever since.

In 1977, I heard a paper delivered by Thomas Berry. The context he outlined also set a direction in my life that has helped me to probe ever deeper the root causes of the kind of dysfunction that plays itself out in the global crises of our times.

Thomas Berry is a renowned historian of world cultures and in his two seminal books, *The Dream of the Earth* and *The Universe Story* , which he co-authored with Brian Swimme, he suggests that the root of our crises is contained with the cosmology that has shaped the total context of western thought. I suggest in this presentation that this cosmology also underpins the world of biotechnology and that it is both flawed and dangerous. I also suggest that a contemporary scientific understanding of the origin, nature, and functions of the cosmos would indicate that biotechnology is itself an extension of the same inadequate worldview and that it is taking us in a direction that is counter to the natural progression of the universe, the earth, and life.

The following chart is my simple attempt to model some of the assumptions that are inherent in our traditional western cosmology. Our origin story, rising several thousand years ago out of the Mediterranean world, provided a context of meaning that attempted to answer the ultimate question of the mystery of existence. In short, it provided a coherent set of meanings upon which the various structures of culture were formed. Some of them are:

- That the divine is totally transcendent to the universe, hence perfect and unchanging.
- The human has a transcendent destiny to be brought into union with the divine, but this union depends on the human transcending the cosmos.
- The cosmos itself does not have this spiritual transcendent destiny. It is a physical, material plane of existence and possesses no inherent spiritual substance.
- The human is free to explore the physical world, analyze its physical energies, and redesign them to bring about some of its original perfection lost after the fall.

Hence, this world view might be described by this simple model:

In this worldview, the ordinary conditions of life are perceived as temporary and abnormal. Thomas Berry suggests that this perception sets the stage for

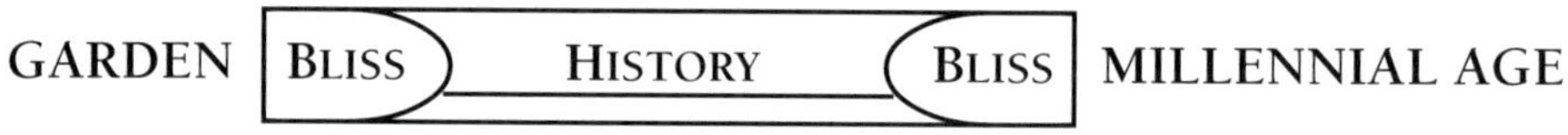

a growing pathological rage within the western psyche. This rage is directed towards the conditions in which life is actually granted to us. Historically, it has made it nearly impossible to develop the inner capacity to live creatively or graciously within the whole fabric of life. Instead, we have resisted all limitations imposed on us as abnormal, as a punishment from which we will one day be liberated. Our inner capacities have been stunted and our total intrusion into the fabric of life, as it has brilliantly evolved, is nearly total.

So it is the scientific story of evolution itself that suggests that our obsession with genetic engineering may well bring about a total undermining of the very life we commit to re-designing. I would like to suggest that as we review the process of evolution, seen now as a total evolution of the inner as well as outer dimensions of the universe, that this context provides an essential correction to the direction that agri-biotechnology is pursuing.

The following overview of the evolution of DNA was prepared by Dr. Lawrence Edwards, PhD, of Genesis Farm.

Evolution of DNA

About 15 billion years ago

The universe flared into existence. At first all was symmetric. Within much less than a millionth of a second the symmetry is broken as the primal four forces emerge. All subsequent relationships will be governed by these four. In particular, the nature of the electromagnetic force is now set. Even though there were no molecules in existence, the laws of chemistry are now in place. So even though no DNA was present, limits were in place on the strength of the hydrogen bonding and therefore on the diameter of the helix.

A billion years later

The universe coalesces into galaxies and stars. The stars live by consuming primal hydrogen and helium and fusing them into new entities — lithium, beryllium, oxygen — all the chemical elements up to iron in weight. The larger stars exhaust their supplies of hydrogen and helium and can no longer sustain their existence as stars. They become supernova and, in that cataclysmic process, fuse to become the heavier chemical elements. Their bodies, rich in chemical elements, are strewn throughout the cosmos.

For billions of years after

Subsequent generations of stars form by gathering the chemically enriched gaseous clouds of hydrogen into themselves, fusing, become supernova, and again distributing more elements into the cosmos.

About 4.6 billion years ago

A large star in our galactic neighborhood became a supernova.

About one hundred million years later

Our sun and solar system formed from the body of this supernova. For several hundred years the planets grew in size by accruing smaller asteroids in often violent collisions. The earth was often molten during this period. During this process the chemical elements born in the star and supernova combine to form simple molecules (e.g., water) and minerals (rocks, stones, etc.).

About 4.1 billion years ago

The great bombardment was over. The solar system reached its present configuration of nine planets. The earth now cooled for the last time, eventually enough so that steam could condense. It rained violently for eons creating the oceans.

About 4 billion years ago

Probably during one or many of those thunderstorms, the first complex molecules were synthesized from the simple molecules and minerals. (No one knows; there are many theories.) Once created, those molecules self-organized themselves and others into creative possibilities. At least one of those possibilities worked. Over the eons those organizational capabilities resulted in the first living organisms. Probably the first genetic capability was through RNA (ribonucleic acid). Later, apparently, DNA (deoxy-ribonucleic acid) proved to be more effective and RNA was then used not for the storage of the genetic information, but only as a "messenger" between DNA and the enzyme production capability. (Again no one is certain of the process in those early years.)

DNA is a chain of four specific nucleic acids. A "word" in the language of DNA consists of a particular sequence of three of these acids. Thus, there are 4x4x4 or 64 possible words in the DNA language. Each word "speaks" of a particular amino acid. So a sentence of words specifies a sequence of amino acids, which is a protein (enzymes are proteins). In all life forms, the same correspondence exists between the DNA words and the particular amino acid. (In many cases there is more than one word possibility for a specific amino acid.)

This was all worked out 4 billion years ago!

Some hundred million years later

Simple bacteria emerge. A bacterium consists of a cell wall surrounding and containing protoplasm, a complex mixture of organic molecules including naked strands of DNA directing the maintenance of the bacterium. The protoplasm connects with the outside environment through the cell wall. Occasionally, the cell clones itself, a process directed by its DNA. Thus, DNA not only remembers how to create a new cell and how to maintain its existence, but also directs the processes.

About 2 billion years ago

In response to the menace of oxygen, a new form emerges, the eukaryote cell, the cell with a nucleus. (Of course, the bacteria live on and prosper without a nucleus.) The cellular DNA is collected and stored in the nucleus as a double helix. This helix unzips during the reproduction process and then each strand duplicates itself in the daughter cells. During this unzipping the DNA is very susceptible to damage. Mutations occur primarily during this time. Apparently this susceptibility to damage is just right: more would result in higher death rates of the daughter cells, less would result in less ability to adapt. (Later, cells developed molecules to "walk" along the DNA helix strand to find and correct errors.)

For the last 2 billion years

DNA has learned, memorized, and directed the processes of life. Changes in the DNA of a particular species have been slow, in earth time. There have been periods of accelerated change, but these periods have still been long — hundreds or thousands of years when compared to human time. All changes were rigorously tested for compatibility with the organism's ecosystem.

15,000 years ago

Humans started consciously changing the DNA of other organisms through horticulture and domestication of animals. Those changes were made much more quickly than normal evolution, but still over many generations. There was not such rigorous testing of the changed organisms and this led to problems in some cases, e.g., exotics taking over an ecosystem. Often the changed organisms are not even independently viable and must be supported by human activities. But, overall, the changes were not large. For example, there was never the mixing of genes between species.

Today

Humans have learned many words and sentences in the DNA language, the means to change sequences within a gene, and the ability to move sequences from one organism to another organism of a different species. Now the time scale of radical evolutionary changes is instantaneous. There is not the time nor the incentive to thoroughly test the new organisms.

There is not even the knowledge of how to test such unknown creatures. While the goal of many of these manipulations is laudable, e.g., the curing of various inherited, debilitating diseases, most are driven by commercial goals. One can imagine taking a certain risk in order to improve the health of a certain segment of the population. We have done this before, e.g., fluoridation of drinking water and vaccinations. But often there have been unforeseen disadvantages to such activities. In any case, much of the genetic manipulation today is for profit. There are few, if any, redeeming qualities except a more efficient product, e.g., a longer lasting tomato.

We do not understand the consequences of genetic manipulations. We are launching yet another massive experiment on ourselves with little understanding of the long-term consequences.

In conclusion

I would suggest that our refusal to live within the limitations of the unity of the whole, which has enabled the elegant miracles of life to unfold, is a dark extension of our mythology. Biotechnology is a commitment to myth. By refusing to acknowledge the superstition implied in our blind adherence to our vision of a world of bliss, we move deeper into a chaos from which life itself may be unable to recover.

That a university would commit itself to direct its young research capacities and its young scientists to such a distorted view of reality is a tragedy, made less understandable with the evidence that science itself holds forth.

REFERENCES

Berry, T. 1988. *The Dream of the Earth*, San Francisco: Sierra Club Books

Berry, T. and B. Swimme. 1992. *The Universe Story*, San Francisco: Harper Collins Publishers

Crop Biotechnology in the Service of Medical and Veterinary Science

CHARLES J. ARNTZEN
Boyce Thompson Institute for Plant Research
Ithaca, N.Y.

A revolution in the basic science related to plants has occurred over the last fifteen years as techniques for genetic engineering have been developed. It is now possible to introduce foreign genes into almost all major crops, resulting in stable genetic transformants that pass the new trait on to their progeny following the principles of Mendelian genetics. Those protocols have led to detailed understanding of the mechanisms by which gene expression is regulated in the different tissues, and at different times, in the plant life cycle. From an applied standpoint, this understanding can be used to cause foreign genes to be expressed only in selected tissues — such as fruits or grains. Plant biotechnology has yielded commercial products in the field of agriculture as plants resistant to insects, viruses, fungi, and herbicides have been created. In addition, foods with modified ripening characteristics are now commercially available.

Beginning in the last few years, various research laboratories have experimented with the use of plants for "biomanufacturing" of specialty products. These approaches utilize transgenic plants created to accumulate high value proteins/enzymes. Some of those studies have explored the production of proteins of potential pharmaceutical value. It has been shown that plants will produce molecules as diverse as human serum albumin (which requires extensive conformation folding) and both serum and secretory antibodies (which require assembly of multiple peptide subunits to produce a functional antibody).

Expression of Antigenic Proteins in Transgenic Plants

It has been well documented that molecular biology has had enormous impact on the fields of immunology and vaccine development. Tools to identify and clone the genes encoding antigenic determinants of infectious agents have been developed and are in wide use. The introduction of the gene encoding the surface antigen of the hepatitis B virus (HBV) in recombinant yeast was the first commercial example of a recombinant subunit vaccine. The HBV vaccine produced by recombinant DNA techniques is extremely safe and effective; however, its high cost has led the Institute of Medicine to list the development of a lower-cost recombinant DNA vaccine as a top priority. It can be readily anticipated that numerous other recombinant subunit vaccines will also become available in the next decade as our understanding of immune recognition of individual proteins grows. While these offer exciting opportunities for disease prevention, dependence upon fermentation technology and protein purification will influence both cost and availability of the new vaccines to the developing world. This led us to explore the use of transgenic plants as a subunit vaccine expression system.

Hepatitis B Surface Antigen Expression in Plants

We began our studies of candidate vaccine expression in transgenic plants using the gene encoding hepatitis B surface antigen (HBsAg). This protein was chosen because the commercially available vaccine and the associated human immune response have been very well characterized, because the structure of the immunogenic form of that protein was known, and because the availability of a cost-effective recombinant HBV vaccine for the developing world is a high priority.

In our initial studies, the gene for HBsAg was introduced into cells of tobacco plants and individual transgenic plants were regenerated. We chose tobacco for the experiments because of its ease of genetic manipulation and abundant literature on controls for gene expression in this facile "model laboratory plant." When transgenic leaf material was extracted, virus-like particles could be recovered. These were characterized and were found to be very similar in structural properties to the recombinant HBsAg, which is formulated in the commercial vaccine produced in yeast cells.

When plant-derived HBsAg was used for parenteral immunization of mice, anti-HBsAg antibodies were recovered that reacted with authentic HBsAg from human serum. This was our first indication that antigenic properties of the protein were maintained in recombinant plants. Subsequently, T-cells were isolated from mice immunized with tobacco-derived HBsAg. When grown in culture, these T-cells could be activated using the commercial vaccine, as well as a synthetic peptide that mimics the "a" epitope determinant of HBsAg. In total, the immunology studies conducted to date show that the recombinant HBsAg recovered from plant cells retains both B- and T-cell epitopes. These studies have demonstrated that plant cells have the capacity to not only synthesize this protein but to allow it to assemble in an immunologically active form.

Vaccines Against Bacterial Diarrheal Disease

Diarrheal disease is the major cause of infant mortality on a worldwide basis. Vaccines to prevent diarrheal disease caused by bacteria or viruses could have a significant impact on human health in the developing world. As my colleagues and I had an interest in testing the oral immunogenicity of recombinant antigens produced in plants, we made an early choice to focus on enteric diseases.

The binding subunit of the heat-labile enterotoxin of *E. coli* (LT-B) was an obvious candidate for evaluation in plant expression systems since it has been extensively characterized in structural and immunological studies. Because it is very similar in structure and immunological properties to the B-subunit of cholera toxin (CT-B), immunization with CT-B leads to cross protection against enterotoxigenic *E. coli* (ETEC). Early field studies of cholera vaccines in Bangladesh showed that CT-B immunization was useful in preventing cholera, although protection was relatively short lived and protective immunity would require frequent boosting.

Our initial studies utilized an expression vector that caused the production of recombinant LT-B in transgenic plants using the native bacterial gene. Characterization of this protein was hampered by the low amount of protein that accumulated in the plants. Subsequently, the bacterial gene was modified to encode a fusion protein, which has a six amino-acid microsomal retention signal at the C-terminus of LT-B. We reasoned that retention of the newly synthesized LT-B in microsomal vesicles of plant cells would allow an increase in the relative concentration of the protein, leading to assembly of the active pentameric form of LT-B. We found that higher amounts of LT-B accumulated, and it could be isolated for characterization. Based on its physical properties and its ability to bind G_{M1} gangliosides, we determined that the LT-B fusion protein assembled into the active oligomeric structure within plant cells. When this protein was partially purified from transgenic plants and given by oral gavage to mice, both serum and secretory antibodies that were specific for LT-B could be recovered from the treated animals. These antibodies were effective in inactivating the *E. coli* toxin when tested using *in vitro* assays as an indicator of protective immunity.

My colleagues and I are continuing our efforts to create transgenic plants that accumulate abundant amounts of recombinant LT-B in edible tissues. We have been guided by the earlier CT-B vaccine field trials that used one milligram of protein per oral dose. By creating a synthetic gene that encodes an LT-B protein of the authentic amino acid sequence, but uses codons which are preferred by plants, we have recently been successful in creating plants that accumulate one milligram of LT-B in a potato that could be consumed raw by a human volunteer. We are cooperating with John Clements of the Tulane Medical School, as well as Michael Levine and Carol Tackett of the Center for Vaccine Development at the University of Maryland, to evaluate this material in human clinical trials for feasibility of use as an oral vaccine.

Vaccines Against Viral Diarrheal Disease

Multiple viruses cause diarrheal disease. Rotavirus may be the principal agent, especially in infants. As we began studies of the potential value of transgenic plants to produce recombinant subunit vaccines, I consulted with Dr. Mary Estes of the Baylor College of Medicine about the potential for a plant-based vaccine for rotavirus. From the data she provided, we concluded that coordinate expression of at least two (and possibly more) rotavirus coat proteins might be needed to cause the assembly of an immunogenic virus-like particle. Since we had not yet evaluated the capacity of transgenic plants to produce immunogenic proteins and because coordinate expression of multiple foreign proteins in plants had not been characterized, we concluded that a simpler system was needed that would provide "proof of principle." For this system we turned to the characterization of a candidate vaccine against Norwalk virus.

Norwalk virus (NV) is a member of the *Caliciviridae* family. It is a causal agent of severe epidemic outbreaks of viral diarrhea. When the gene encoding the single capsid protein of Norwalk virus was expressed in insect cells, virus-like particles (VLPs) could be isolated. These particles were also found to be effective in causing oral immunization of mice, resulting in production of both serum and antibody specific for the capsid protein. Unfortunately, there is no animal model to demonstrate protective immunity of vaccines against the Norwalk virus; those determinations will require human clinical trials.

Extracts of plants expressing the NV capsid protein were found to contain VLPs that mimicked the structural properties of VLPs recovered from insect cells. When these were partially purified and given to mice by gavage, both serum and secretory antibodies to the NV capsid protein were produced.

Oral Immunization Using Edible Plant Tissues

Although our initial subunit vaccine experiments used tobacco as a test system, our goal was to produce candidate vaccines in edible plant tissue to evaluate the potential for immunization simply by eating the tissue. Because the oral immunogenicity of plant-derived antigens had been demonstrated by the oral gavage studies, we developed protocols for genetic transformation of potato using genetic regulatory elements that cause the accumulation of the desired recombinant protein in the potato tuber tissues. We chose potatoes for two reasons. First, we could recover transgenic plants and grow them to maturity in greenhouses in a relatively short period (about three to four months). Secondly, mice will readily eat raw potatoes.

Transgenic plant material has been generated that contains both recombinant LT-B and the recombinant Norwalk virus capsid protein. The proteins assembled into the appropriate structures (LT-B pentamers that bound G_{M1} gangliosides, or VLPs for the NV capsid protein). Potatoes were peeled, and five gram samples were fed to mice on a schedule that was analogous to the previous experiments using gavage for oral immunization. In both cases, the mice produced serum and secretory antibodies against the recombinant protein in the potato eaten by the mice as food. These studies provide "proof of concept" for edible vaccines.

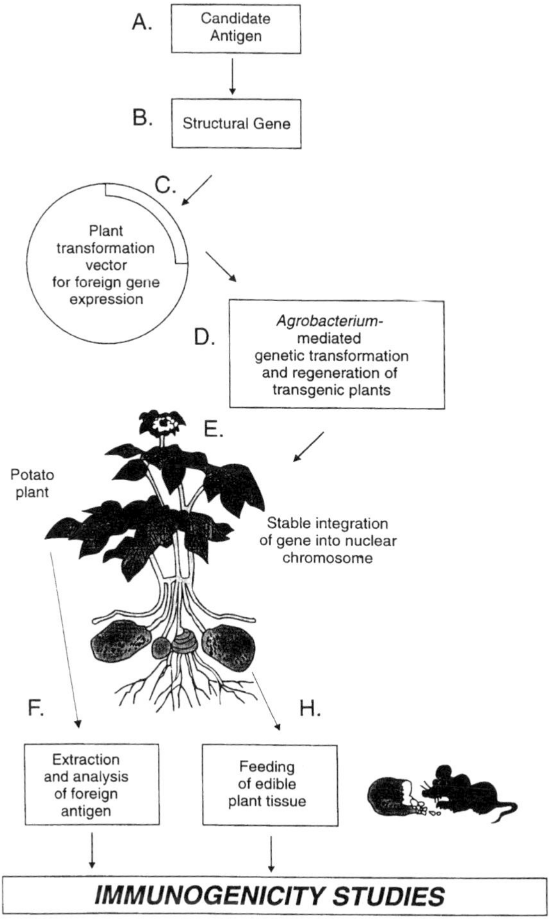

Figure 1. A schematic diagram showing the steps involved in creating transgenic plants expressing immunogenic proteins. **A.** Based upon earlier immunology studies, a candidate antigen with known or potential value as a subunit vaccine is identified. **B.** The structural gene encoding the antigen of interest is obtained or isolated. **C.** The gene is ligated into a plant transformation vector which contains plant-specific promoter and terminator sequences to regulate gene expression. **D.** The natural gene transfer system of *Agrobacterium tumefaciens* is usually utilized to mobilize the genetic construct into the chromosomes of plant cells. **E.** Plants are regenerated from transformed cells that contain a stably integrated gene encoding the desired antigen. **F.** and **G.** Immunogenicity of the plant-expressed antigen is determined by either extracting the foreign protein and delivering it by parenteral or oral routes, or simply by feeding a plant sample as food.

Vaccines for Animal Diseases

Oral vaccines can provide efficient and humane strategies for disease prevention in production and companion animals, as well as feral populations. The above-mentioned "proof of concept" for edible vaccines suggests that the design of subunit vaccines in feed crops is a viable strategy. Seed-specific genetic regulatory elements are known for crops such as corn and soybeans, and stability of foreign proteins in dried seeds has been demonstrated.

Recently, transgenic plants have been generated that expressed the gene encoding the glycoprotein (G-protein), which coats the outer surface of the rabies virus. Although the immunogenicity of these materials has yet to be reported, it is encouraging to note that bait containing some G-protein produced in a more traditional in-vitro system was effective in orally immunizing raccoons, providing protection against "street virus" challenge.

Remaining Questions/Opportunities for Edible Vaccines in Transgenic Plants

The research conducted to date has demonstrated that transgenic plants have the capacity to synthesize and accumulate subunit antigenic proteins that retain immunological properties of their native counterparts. In the case of the HBsAg and the NV-capsid protein, virus-like particles accumulate; this may be very significant because the particulate form of the proteins is likely to be important in determining immunogenic properties. It is likely that particulate structures will have greater oral immunogenicity than soluble proteins.

Studies not yet conducted will involve the evaluation of dosage requirements for plant-delivered vaccines. Successful experiments conducted thus far have used proteins from two enteric organisms. It will be necessary to determine if other proteins, which may not be normally transmitted via the oral route, will be as effective in inducing an oral immune response. We are optimistic, however, because plant cells represent a natural bioencapsulation system with surrounding layers of cell wall, cell membrane, and (in some cases) internal membrane compartments to encapsulate and thereby protect the desired subunit protein from digestive degradation. It remains to be determined if the release of the desired protein in the gut acts as a "slow release" as the plant cells are degraded in the normal digestive process. If so, dosage levels may have to be adjusted accordingly.

A potential major advantage of recombinant plants for vaccine production is the possibility that multi-subunit vaccines, including an oral adjuvant such as CT or LT (or derivatives thereof) could all be produced in a single plant. There is no theoretical limit to the number of different genes that could be introduced into a single plant species. Plant tissue could, therefore, contain multiple antigens in one delivery system. In addition, this could circumvent the need for a cold chain in vaccine delivery when produced and utilized in developing countries.

It is well recognized that most food proteins do not trigger an immune response; in general this is due to the induction of a state of immune tolerance. It will be necessary to determine if food-based vaccines will also induce oral tolerance to the desired antigen. If so, controlled use and dosage will be a requirement for "edible vaccines." This should not be an insurmountable obstacle, however, as only a small amount of plant material would need to be propagated for wide-scale vaccine delivery and its distribution could become a component of public health care systems.

The type of plant material that would best serve as an "edible vaccine" is yet to be determined. For human vaccines, our own research team has focused on the use of bananas. This crop has three major attributes: it is grown in almost all tropical or sub-tropical developing countries throughout the world; the food is eaten uncooked (which would avoid denaturation of subunit proteins); and bananas are a food that is widely consumed by infants and children. At the present time, we have developed a methodology for the genetic transformation of bananas and are cloning fruit-specific genetic regulatory elements, which we believe will cause the tissue-specific production of the desired candidate vaccine in the developing fruit. The primary disadvantage of bananas is one of technical limitations during feasibility stages of research. This is due to the fact that the time from genetic transformation until harvest and evaluation of the fruit is relatively long. We anticipate that the time period will be at least two years. Our first transgenic banana plants containing genes encoding candidate vaccines are still in the seedling stage.

For animal vaccines, a variety of grain crops (such as soybeans or corn) represent excellent possibilities for vaccine delivery. Transformation techniques for those crops are known, as are strategies for causing seed-specific gene expression.

In summary, the expression of vaccines in plants may be the first large-scale example of production of high-value pharmaceuticals in transgenic crops.

SELECTED READINGS

Clemens, J.D., D.A. Sack, J.R. Harris, J. Chakraborty, P.K. Neogy, B. Stanton, N. Huda, M.U. Khan, B.A. Kay, M.R. Khan, M. Ansaruzzaman, M. Yunus, M.R. Rao, A-M. Svennerholm, and J. Holmgren. 1988. Cross-protection by B subunit-whole cell cholera vaccine against diarrhea associated with heat-labile toxin-producing enterotoxigenic *Escherichia coli*: Results of a large-scale field trial. *J. Infectious Diseases*. 158:372-377

Clemens, J.D., D.A. Sack, J.R. Harris, F. van Loon, J. Chakraborty, F. Ahmed, M.R. Rao, M.R. Khan, M. Yunus, N. Huda, B.F. Stanton, B.A. Kay, S. Walter, R. Eeckels, A-M. Svennerholm, and J. Holmgren. 1990. Field trial of oral cholera vaccines in Bangladesh: Results from three-year follow-up. *Lancet*. 335:270-273

Haq, T.H., H.S. Mason, J.D. Clements, and C.J. Arntzen. 1995. Production of an orally immunogenic bacterial protein in transgenic plants; Proof of concept of edible vaccines. *Science.* 268:714-716

Holmgren, J., N. Lycke, and C. Czerkinsky. 1993. Cholera toxin and cholera B subunit as oral-mucosal adjuvant and antigen vector systems. *Vaccine.* 11:1179-1184

Jiang, X., M. Wang, D.Y. Graham, and M.K. Estes. 1992. Expression, self-assembly, and antigenicity of the Norwalk virus capsid protein. *J. Virol.* 66:6527-6532

Lyons, P.C., G.D. May, H.S. Mason, and C.J. Arntzen. 1996. Production of protein pharmaceuticals in transgenic plants. *Pharm. News.* 3(3): 7-12

McGarvey, P.B., J. Hammond, M.M. Dienelt, D.C. Hooper, Z.F. Fu, B. Dietzschold, H. Koprowski, and F.H. Michaels. 1995. Expression of the rabies virus glycoprotein in transgenic tomatoes. *Bio/Technology.* 13: 1484-1487

Mason, H.S. and C.J. Arntzen. 1995. Transgenic plants as vaccine production systems. *Trends in Biotechnology.* 13:388-392

Mason, H.S., J.M. Ball, J-J. Shi, X. Jiang, M.K. Estes, and C.J. Arntzen. 1996. Expression of Norwalk virus capsid protein in transgenic tobacco and potato and its oral immunogenicity in mice. *Proc. Natl. Acad. Sci. USA.* 93:5335-5340

Mitchell, V.S., N.M. Philipose, and J.P. Sanford, Eds. 1993. *The Children's Vaccine Initiative.* Washington, D.C.: National Academy Press

Thanavala, Y., Y.-F. Yang, P. Lyons, H.S. Mason, and C.J. Arntzen. 1995. Immunogenicity of transgenic plant-derived hepatitis B surface antigen. *Proc. Natl. Acad. Sci. USA.* 92:3358-3361

Sludge, States Rights and Success

KEN EVANS
President Arizona Farm Bureau
Phoenix, Ariz.

As I will describe later, we have sophisticated technology to handle agricultural applications with sludge. With this technology you would expect government regulators to be supporting our efforts. Not so. Our biggest enemy and clearly our most serious obstacle, whether in wildlife preservation or quality beneficial bioremediation, is the very government reportedly there to protect us. The Federal government's weakening of regulatory processes at the state level has created a time bomb — an environmental disaster just waiting to happen or be reported. I do not want to be a continuing part of an industry "in retreat" from prudent regulation. My only forum for available protest is to retreat from that management practice until prudence returns.

What a wondrous time we live in. At no time have the twin spears of technology and communications had such impact and influence on people's lives. Over the years, I have traveled a great deal and have seen much of what the world has to offer. On my travels, one observation stands out above the rest. From New York to Tokyo, from Cartegenia to Boise, from Tel Aviv to Ciudad Obregon, from Mexico City to Kansas City — no matter how small the cardboard hut or how exquisite the mansion, no matter how beautiful the home or how small the apartment, most all have televisions. Whether in the streets of Mexico City or the ghettos of Los Angeles, the most affluent neighborhood in Scottsdale or the middle class suburbs of Atlanta, being able to see the news and the latest fashions instantly has produced a silent but powerful revolution in thinking throughout the world. Under all of the extremes of social and economic class differences, designer sneakers and jeans are there . . . and there in abundance. Kids may not have enough to eat but they know what's "in" and they wear it. They see political and economic and social conditions elsewhere and they want the best of what they see. This worldwide communication explosion has brought images of Robin Leach's "Lifestyles of the Rich and Famous" into people's homes, dreams into their eyes, hope into their hearts, and an insatiable hunger to their belly. The bringing of news and current affairs and entertainment into the homes and lives of earth's inhabitants will one day be recognized as the most significant political, social, and educational event in recorded history. Televisions were the powerful fans that produced the winds of change around the globe. But television will be to the telecommunication, tele-information and tele-management tools of the 21st century what slide rules were to computers.

I will now describe the advanced technology used for sludge application at my farm.

In the early morning dawn, as the spine-tingling scream of the 903 diesel engine grew close, it became apparent to the college visitors that the machine was using all of its power to pull the seven tine, deep chisel plow. But it wasn't the display of raw power that caught their attention. Nor was it the precision driving that had kept the tractor on target within six inches across a half mile field — 2,600 feet. Neither was it the high-tech instrumentation that had regulated the application rate to within 0.03 percent of the target rate of 48,450 gallons per acre. It wasn't the cab's air conditioned, stereo tape deck, CD ROM dust-free environment or its computer instrumentation. What impressed the visitors to this 12,000-acre southwestern family farm was their discovery, as the tractor turned, that there was no driver. The 300 horsepower sludge injection tractor was controlled remotely by a portable computer sitting on the seat of my Chevy pickup.

The use of biosolids, as better quality sludge is now called, has been a scientifically based management tool on our ranch for the past 18 years now, but that will likely come to an end as a result of government abandonment of reasonable standards for the industry. When was the last time you heard a farmer say he wanted more government regulation? More on that later. Over the years, we have carefully applied and tracked over 3,000,000 metric tons of municipal biosolids. We have actually raised the elevation of our 22 square mile ranch over three inches.

What in the world do biosolids have to do with biotechnology? Well, as Peter Day, Director of the AgBiotech Center at Rutgers University, told this group in 1994, one of the most underdeveloped opportunities for agricultural biotechnology is in the area of bioremediation. For those who have visited our farm, however, this is not the typical sludge dump-and-run operation you see all across America. In addition to the high-tech application equipment, which my family has pioneered, we have also developed high-tech, biotechnology methodologies to monitor and track nutrients, pathogens, and heavy metals. As an example, we have pioneered the use of DNA/PCR gene probe testing to provide us with analyses of Escherichia coli pathogen levels in the biosolids prior to application. To give you a scoping perspective, we have run more than 22,000 pathogen tests this year. Additionally, we use a Strontium 90 CPN Neutron Probe to give us instant measures of heavy metal concentrations in the biosolids so we can track those as well. In the last six months we have diverted nearly 1,200 truckloads of biosolids to a landfill because it did not meet our quality standards. Additionally, our reporting is done electronically using an instantaneous and accurate reporting protocol.

Today, I have been asked to focus on regulatory and social changes affecting Agricultural Biotechnology in the Environmental and Energy Sector. Specifically, where we are, where we're going and how we're going to get there. The biotechnology sector is about where the budding computer industry was in 1976. We are facing enormous changes in the agricultural sector over the next few decades. Let's discuss a few of the changes that will most affect us.

1. Farmers of the future will produce more food and fiber, and more than just food and fiber.

Agricultural biotechnology will affect every single farmer and rancher. As the emphasis shifts from yield per acre to yield per unit of resource expended delivering products to the end-consumer, biotechnology will play an increasingly important role. Additionally, crops that will be engineered specifically to yield feedstocks for the pharmaceutical, energy, and industrial sector will dot 21st century farms. Growing fuel and cosmetics will become as important as growing feed and cotton.

2. Production of industrial and chemical feedstocks will be a major source of revenue, not a novelty.

As the percent of GDI spent on food and clothing continues to decline along with the demise of government subsidies, a new generation of farmers will emerge. The appropriate infrastructure must be embraced in order for this to occur globally.

3. Tomorrow's farmers will be as comfortable on the internet as yesterday's were on a tractor, and as comfortable using a computer as their dad was using a shovel.

The internet has evolved from an information gold mine to an information glut, to a management and marketing tool. This evolution, as discussed earlier, will touch every aspect of our lives. Smart computers, tied to worldwide information and data, will allow individual producers to make better decisions. We must develop an infrastructure that allows open and free movement of information between private business and government, between the regulated and the regulator. We must give up old crutches and embrace new concepts of community welfare, individual accountability, and shared responsibility for the public and environmental well-being of this space ship we call Earth.

4. Conservation will become an established philosophy religiously practiced — not a mandated farm practice.

The worldwide environmental movement and community sensitivity about environmental and wildlife protection, coupled with global access to what we do, or don't do, on our farms, will force farmers and ranchers to modify their farming philosophy to take into account public attitudes about the impact of farm practices on the environment.

5. A new alliance will emerge between land grant universities and commercial researchers and farmers, end-users, and true environmentalists that will carry us well into the next century.

The next decade will produce a blurring of the traditional structural lines in business and society. Responsibility for the administration of social goals will continue to evolve away from government and toward a shared public-private partnership. From education to medical care, from basic research to child care, from prisons to the airwaves, a blurring between private enterprise and government regulation will occur at an ever escalating pace. Even the lines between labor and management will become hard to distinguish. Banking, insurance, and marketing will blur until they cease to exist as discernible institutions and focus instead on function and vision. To succeed, we must identify, develop, and cultivate those new alliances that share our common vision. We must also tell our story. We must spend more of our resources educating the public and ourselves. The public's right-to-know and their access to huge amounts of information regarding every aspect of business and the environment will force us to be more open and forthright in business practices and government operation. We must address phobias, fears, and suspicion by the public and within our own industry. We must be willing to change.

6. Farmers who succeed will need to expand into providing for other societal needs such as recreation, wildlife management, environmental management, etc.

The most recent USDA report revealed that the 1.5 million small-and medium-size American farmers received nine-tenths of their disposable income from off-farm sources, with even the largest producers receiving more than half of their disposable income from off-farm sources, as well. As a further indication of this trend, three State Farm Bureau Presidents make more money on wildlife and recreation than on the cattle or crops on their ranches. American farmers in the 21st Century will have to learn to farm for the maximum benefit of the entire ecosystem we call Planet Earth versus mining the land or farming directed by government programs. Providing recycling for such waste products as biosolids and the beneficial reuse of our dwindling resources will become more widespread. Biotechnology will play a critical part in accomplishing this goal.

7. We must do a better job of telling our story.

We have accepted and embraced proven, as well as emerging, technology. The public does not understand what we are doing or why. In response to Caron Chess, yes there is a segment in society that has a higher latent trust factor than the "white-coated scientist." They are the family farmers. There is a difference between having information versus having understanding. To prevent special interest groups of political activists and environmental extremists from misusing their access to massive amounts of data to create mass hysteria, we must empower the average citizen with access to and understanding of this ocean of information. We have heard much about the shortcomings of the nuclear industry or the chemical industry or the plastics industry. Perhaps the greatest weakness and most damning deficiency of modern society is our illiteracy when it comes to understanding risk. As a people, we expect the government to give us freedom from all risks, even those we can't define. While one-third of the world is hungry, we demand: convenience without cost, pleasure without pain, thrills without danger, recreation without work, service without sacrifice, freedom without price, glory without honor, products without waste, energy without pollution — any at all. We expect to have toys that do not break, cars that don't wreck, trains and planes that don't crash — ever! We want to gamble and win, but do not accept the risk of losing. We even want tans without sunshine and health with out prudence. We fantasize about nature being in balance and then bankrupt ourselves and squander our children's future trying to restore a balance to nature . . . a balance that never was nor ever, ever will be. To survive we must educate an entire generation about the concept and cost of risk avoidance. Risk cannot be eliminated. Risk can be reduced — at a cost. Cost benefit analysis must be a core part of education.

These changes are occurring, with all of their attendant challenges and opportunities. I, like you, have a hard time visualizing what the 21st Century farm community will look like. But this much I do know. Telecommunication assisted, real time, graphically enhanced technology will put substantial regulatory and public pressure on every aspect of business and education including the development and commercialization of biotechnology. In spite of that, biotechnology will play a significant role in the future of farming.

Innovation, Industrial Development and the Regulation of Biotechnology

MARTINE KRAUS
Center for the Study of Law and Society
University of California-Berkeley
Berkeley, Calif.

We have heard many bad things about the U.S. government and government regulation; people have argued that there is either too little or too much regulation in U.S. agriculture. I am going to say a few good things about the U.S. biotechnology regulatory regime.

As you all know, regulation serves a role in safeguarding the environment and assuring food safety in the agricultural industry, and product safety, efficacy, and the consistency of the manufacturing process in the pharmaceutical industry. In both the food and pharmaceutical industries, regulation and government approval raise consumer confidence and provide companies with a powerful marketing tool — the label that says "USDA" or "FDA approved." The industry wants regulation. One only has to recall the case when Richard Godown from the Biotechnology Industry Organization and Rebecca Goldburg from the Environmental Defense Fund went jointly to the Council on Competitiveness asking the council to please not deregulate too much.

On the other hand, there is legitimate concern about regulation stifling innovation, the economic implications of regulation, and questions of international competitiveness in comparison to Europe and Japan. Over the years, the EPA has been criticized for not issuing its policy guidelines on time; the USDA for over-regulating or deregulating the wrong crops; the FDA for its long review times and requirements that were supposedly greater and more burdensome than their European counterparts.

Hence, the desire for regulation and the fear of regulation are hostages of each other. To have the seal of approval on your product, you need regulation. And regulation is always slower than no regulation.

In the case of biotechnology, the relationship between regulation and economic development has been at the center of all regulatory debates, I believe more so than in any other industry. There are three reasons for this phenomenon. First, the technology is thought to hold enormous economic promise and social benefit. Second, the industry is composed of many small start-up companies with limited resources to devote to regulatory affairs. Third, we are dealing with a sophisticated industry whose players have been, from its origin, very involved in regulatory matters. It was the scientists who came to Asilomar in the 1970s, many of whom subsequently became the founders of biotechnology start-up companies and the industry as a whole.

Many years later, the major question is: *Has regulation negatively or positively affected the development of the U.S. biotechnology industry?* I will speak to this subject in comparison to Europe. I will primarily address the regulation of plant biotechnology, but will also mention the regulation of the medical biotechnology industry because the pharmaceutical industry is prominent on this year's program. I will begin with my conclusions.

Conclusions

- *The U.S. plant biotechnology industry has been positively affected by domestic regulation and finds itself at a competitive regulatory advantage to its European counterpart.*
- *The industry has, however, been negatively affected by European regulation, in that companies hesitate to develop export crops for the European market or make investments in Europe because of perceived regulatory uncertainty.*
- *The U.S. medical biotechnology industry is domestically subject to very burdensome and stringent regulatory requirements. The industry is, however, not more disadvantaged than its European counterpart, because both regulatory regimes are costly and provide few incentives for the regulator not to err on the side of caution.*

Those conclusions, as well as the subsequent arguments, are based on studies of companies with similar operations in Europe and the United States, of which I will give you an example. I call it: A Tale of Two Companies.

A TALE OF TWO COMPANIES

This is the example of a U.S. and a German based company, both of which developed a virus resistant crop, a squash, and a sugarbeet. Each crop raised very similar regulatory concerns in each country, i.e. questions of gene transfer and cross-compatibility with wild relatives. In order to gain a permit for field testing the crop, the U.S. based company faced a one-step regulatory process involving primarily the USDA-APHIS. The German based company instead faced a three-step regulatory process. The company first needed to submit its application to the national authority. The national authority in-turn sent a summary of the company's dossier to the European Commission, that, in-turn, sent the summary dossier to all member states of the European Union.

As far as the applications are concerned, the U.S. application consisted of 21 pages: seven pages on host and recipient organisms, vectors, and the genetically modified crop; four pages on the purpose of the field test, methods of data collection and harvest procedures; and one paragraph on the location and supervisory personnel. The remaining pages were graphs to support the above. The German application consisted of 100 pages: 60 pages on host and recipient organisms, vectors, and the genetically modified crop; 30 pages on the purpose of the field test and environmental risk; and 10 pages on company personnel and the supervision of the test.

Monitoring requirements in the U.S. call for only an annual field trial report, while the German company must issue a mid-trial and final environmental assessment report. In a 1995 article in *Bio/Technology*, Margaret Mellon and Jane Rissler showed that the U.S. monitoring requirements are inconsistently implemented by the USDA, while in Germany the implementation is consistent.

The initial review time for the U.S. application was 118 days. This number was reduced as the agency gained experience with the company's product and subsequent field trials were approved much more rapidly. In Germany, the approval took 175 days, followed by an appeal to court that added 32 days to the overall approval time.

The overall cost was greater in Europe than in the U.S. The estimated cost of company regulatory affairs time in the U.S. was $5,000, while in Germany it was approximately $100,000. An application fee does not exist in the U.S. but is in the range of $7,500 to $45,000 in Germany, dependent on the time and effort it takes the agency to review the application. A legal cost that is not an issue in the U.S. was estimated by the German company to be approximately $100,000.

This company comparison illustrates what I call the American regulatory advantage in plant biotechnology. The American regulatory advantage in plant biotechnology is two-fold. It consists of the regulatory regime per se, and companies' regulatory proficiency.

The Regulatory Regime

Regulatory Structure: The U.S. regulatory regime is much more centralized than its European counterpart and presents fewer regulatory hurdles to the company.

Regulatory Requirements: The U.S. regulatory requirements are far less burdensome than their European counterpart where more data regarding the environmental safety of the field test, the specifications of release conditions, and the monitoring and control of test sites are required.

Stringency: U.S. regulatory requirements are less stringent than their European counterpart.

Regulatory Style: The U.S. regulatory agencies are less legalistic than their European counterparts, clearer in their requirements, more cooperative, and less adversarial.

Review Times: Review times are consistently faster in the U.S. This is more so since the introduction of the notification process for well-characterized crops.

Regulatory Certainty: In comparison to Europe, the U.S. issued its guidelines for field-testing genetically modified crops and the commercialization of transgenic foods much earlier.

Statutory Flexibility: In the U.S., a single agency can adapt to technical progress given requirements and policies. In Europe instead such decision has to be taken by the European Commission and in part with the agreement of the European Parliament.

Other speakers may discuss whether these U.S. regulatory advantages come at a cost of increased environmental risk and reduced safety. I believe that one can have an efficient regulatory regime that safeguards the environment.

The second important part of the American regulatory advantage is the companies' regulatory proficiency or their capacity to respond to regulatory challenges.

Company Regulatory Proficiency

Company Organization: U.S. biotechnology companies are staffed with a director for regulatory affairs who is actively involved in product development. In Europe it is often the lead scientist who interacts with the regulator.

Company Experience: U.S. biotechnology companies are often experienced in dealing with environmental regulation, as often they are managed and/or staffed by former employees of the chemical industry (e.g. Mycogen's CEO formerly worked at Monsanto) or former employees of the EPA. Many European companies have never been regulated on environmental grounds. European regulators rarely move into the private sector.

Status of Regulation: Many U.S. companies consider regulation an integral part of product development. Many European companies consider regulation after they have developed their product.

Company Ingenuity: Some of the U.S. biotechnology companies established relatively early proved to be extremely innovative and imaginative in their response to regulatory challenges. European companies have not shown such ingenuity.

Following the above, it is tempting to conclude that the success of the U.S. agricultural biotechnology industry, in comparison to its European counterpart, is in major part the result of the regulatory proficiency of both the industry and government.

I conclude, however, that this is not entirely so. Regulation is only one factor in a complex innovation system. While regulation is critical at certain stages of product development, it rarely determines whether a company is founded, an entrepreneur realizes his or her plans, a product gets developed, or a technology is adopted. Hence, it does not have a direct impact on innovation per se, but is always reactive. Only in situations where there is extreme regulatory uncertainty can regulation seriously affect the development of an entire industry.

Why have the same set of techniques generated so much more industrial activity in the U.S. than in Europe (measured in terms of the number of field tests)? I suggest the difference goes beyond regulation to the industrial organization in place at the time of the introduction of the technology.

Industrial Organization

When biotechnology was introduced, there was an established European chemical industry (e.g. Hoechst, Bayer, BASF) that was dominated by chemists who were reluctant to enter biotechnology and had historically never been in the seed business. On the other hand, the European seed industry consisted of small- and medium-sized companies that were mostly family-owned and very traditional in nature. They were slow to innovate, reluctant to go high-tech, and had limited resources.

In contrast, the United States biotechnology industry consisted of start-up companies and established chemical companies. The start-up companies were very innovative and open to risk-taking, specialized in biotechnology techniques, and were staffed by university researchers and molecular biologists with close ties to the research community and the land grant institutions. Those companies were highly dependent on product approval because of the need to demonstrate to the venture capital community that they were worthy of funding. In most cases, the established chemical companies (Monsanto, Dow etc.) were active in crop protection and, in some cases, the seed business. In the 1970s, many U.S. chemical and petro-chemical companies had bought into seeds. The large chemical corporations also benefitted from the close proximity of innovative start-ups. Finally, the U.S. industry consisted of the world's largest seed company (Pioneer) with large resources. This U.S. industrial organization favored it over its European counterparts.

In conclusion, I wish to emphasize the following points:

- *Regulation plays an important role in assuring consumer confidence. This is particularly so in a new industry such as biotechnology.*
- *A good regulatory framework, such as the U.S. framework for the regulation of plant biotechnology, provides a support structure to industrial development but is not critical to innovation.*
- *The largest responsibility for innovation, industrial development, and the success of the regulatory process lies with individual companies, their organization, and the industry as a whole.*

Constructing Food for Shareholder Value

BREWSTER KNEEN
The Rams Horn
Mission, British Columbia, Canada

> *The Food Industry Workshop: "Evaluate the pathways to be taken by new agricultural biotechnology food products; issues of communication, regulation, and concern over something so 'personal' as food; and ways in which the production of new food crops might impact the structure of the agricultural industry and the food industry itself."*

This is one way to frame the discussion, but we must recognize that the title itself — food industry — is a cultural expression. It is a way of seeing the world. It also implies a monoculture: there is one correct way to view the world, including food and agriculture. This is the way the wealthy industrialized societies of the North view the world out of their historically and culturally distinct experience. But what about other perspectives and experiences? A woman's perspective, or that of a laid off worker? Or the perspective of a lesser developed country's subsistence farmer? Or an aboriginal in Costa Rica, or in the southwest or north of this continent?

Are we even aware of the biases and assumptions that we take on when we use the cultural expression "food industry" and do we, or are we willing, to take account of how others might view these issues?

I am not castigating the NABC, because when I raise these alternatives, I must say that the NABC looks like the most thoughtful and sincere discussion on biotechnology going. Look, for comparison, at the hype surrounding the 1996 "Agbiotech International Conference" being held in Saskatoon. "New Technologies! New Marketing Partners! New Opportunities!" is the conference slogan. Complete with exclamation marks. In keeping with their slogan, I can find no indication in their program that there will be any critical discussion of anything. (Not that this surprises me, given who has organized the event.)

Or look at the program for Bio '96 in Philadelphia. As far as I can see from the program, there is one workshop, out of about 80, that deals with ethics, titled "Characterizing & Addressing Ethical Issues." The information package states "The Bioethics Committee and industry representatives will discuss case studies where research and product development yield ethical implications, and will participate in analysis of ethical implications of case studies."

The language boggles my mind: what is meant by "yield ethical implications"?

At first I thought there had been a mistake when I saw that I was down to speak in the Economic and Structural Issues session, rather than yesterday's session on Social Issues, Regulations and Ethics. But as I pondered it, I realized that this is indeed the right slot for me. When I studied theology, ethics was my major interest. The subject was referred to as Christian ethics or social ethics, and the focus was largely political and social, not individual and personal. Since then, I have found myself increasingly focused on economics and its institutions and structures — as well as the values and assumptions on which they are based. Meanwhile, it seems to me that 'ethics' has been increasingly defined in individualistic terms as issues of personal choice. The rising field of medical ethics and bioethics is a good illustration of this, where individual choice reigns supreme and social good does not seem to be even a recognizable category, much less an ethical framework.

A good example of the application of this individualized ethics is the V-Chip, described by *New Scientist* magazine as a way "to banish TV sex and violence from American homes." This bit of technology seems to be based on the assumption that there is no way, or desire, to get rid of TV sex and violence, along with an acknowledgment of negative effects on children. An obvious social problem of rather substantial ethical significance is reduced to a matter of individual parental option.

And while it may strike you as stretching the point, it seems to me that the labeling of genetically engineered foods is being approached in much the same way. There appear to be no questions raised either about the production of violence on TV or about the production of genetically engineered food. In the case of food, the biotechnology industry promotes the position that it would be absurd, unworkable, and costly to label genetically engineered food. In a 1993 interview, an FDA spokesperson said: "It would not be merely a matter of putting a sticker on a tomato or a banana. Producers would have to segregate the genetically engineered foods from other varieties. Does the label have to follow the food processing chain? It would increase the cost of these foods to consumers and would disrupt our complex food distribution system."[1]

[1] James Maryanski, biotechnology coordinator, Center for Food Safety and Applied Nutrition, FDA, in a 1993 interview by FDA consumer writer Mary Alice Sudduth.

Monsanto's 1995 kit for the American Dietetic Association[2] takes the same line: "Some consumers believe that foods derived from new plant varieties [notice the slick linguistic trick of referring to "new plant varieties"] should be labeled as such. These individuals [only "individuals," of course] base their position on the consumer's 'right to know' the food's method of production. Practical considerations make such a position difficult to implement . . . and would result in increased costs to consumers with no benefits." Except, of course, their ability to reject Monsanto's biotechnological products.

But if there is a problem of informing the public about what they are expected to purchase and eat, then maybe the production of these foods in the first place is the real problem, and not labeling at all.

Since this session is on the subject of economic and structural issues, I must ask why it is that our culture is so loath to examine the structures of business. For myself, I find it necessary to focus on structures and institutions in order to properly frame the ethical and social issues, and I seldom use the language I was once trained to use. In fact, I was a little appalled when I pulled some old textbooks off my shelf and looked at them from my current perspective. They were terribly inadequate in the way they framed the questions and in their cultural bias — typical, I am afraid, of the chauvinism and imperialism of American culture at the time. But is it any different today?

The processes and products of agricultural biotechnology, as well as their human counterparts, are expressions of a particular historic culture: the culture of Enlightenment, Industrial Revolution and Reductionism Science. Biotechnology is an artifact of this culture. It is no more a universal outcome of evolution than our notion of science itself.

In fact, we might describe biotechnology as an apocalyptic remnant of the 19th and 20th century idea of Progress and Development — sustainable or not. Just recall the language you see in virtually every piece of literature on biotechnology: "improved" seeds, "superior" genes, "best" traits, and so on. Meanwhile, fewer and fewer people believe in the myth of progress — and those who still do are generally of the white northern business class — a rather small elite of the global population whose own children appear to be rejecting the idea.

Stripped of its emotive language, biotechnology would look like Swiss cheese — or the brain of a cow suffering from BSE!

---

[2] "Plant Biotechnology — Harvesting Solutions for Tomorrow's World", produced by Monsanto Company in cooperation with The American Dietetic Association, 1995.

On the basis of its belief in a deterministic understanding of evolutionary progress, it is not surprising that our culture has taught us that there is only one valid epistemology, only one way to know about life and the world. We might well apply the term 'monoculture' both to the knowledge system of our rationalist, industrial culture, and to its practices, for example agriculture. Our notion of science and technology is, in fact, based on an epistemological monoculture. One might well add that sexism, racism, and nationalism are also expressions of monoculture. As in a field of hybrid corn, variation and differentiation are not welcome! They interfere both with the images we have come to accept as normative and with the production process. The model is still the production line. (You should see the difference in my garden and my daughter's — I have a hard time not planting in tidy rows, while hers is helter-skelter — herbs, flowers, and vegetables in glorious confusion.)

Lest you think I am being extreme, let me remind you that not very long ago the term "junk DNA" was used to describe what could not be explained: because it could not be explained, it was junk. Prions did not exist very long ago either, much less mutant prions that seem to cross species barriers on their own. And an article on the genetic blueprinting of yeast in the April 27, 1996 issue of *New Scientist* made the interesting comment that, "Sequencing the yeast genome has revealed a vast *terra incognita*. Biologists have no clue as to the function of 40 percent of the genes they have identified. Half of these enigmatic genes have DNA sequences similar to other, equally puzzling genes in fruit flies, mice, or other organisms, but half have never been seen before. Researchers have dubbed these genes 'orphans' because no one knows which gene families they belong to."

I suppose the common response is that we will soon fill in those blanks and complete our knowledge. But will this be the case, or only the case as we wish it to be? Unfortunately, we are all too likely to simply ignore (or eliminate or "disappear") what does not fit, or, for that matter, other ways of knowing.

I do not say this with malice. I, like most of you, grew up in an era and a culture of imperialism — only we called it "development". In this culture it has been assumed that everyone wants to, and could, become like us. And, of course, it has been assumed that we know how to make this happen. Agricultural biotechnology, as a reincarnation of the Green Revolution, and based on the same neo-Malthusian assumptions, is simply the latest artifact to be devised in pursuit of this goal. Monoculture is a perfectly reasonable phenomenon in this context. There is one goal, and one way to achieve this goal. Certainly the propaganda of industrial biotechnology expresses this monoculture.

But I have been taught by others, as well as by my own experience, which I refuse to invalidate, that there are always more ways than one to know anything and to do anything. It all depends on perspective, experience, culture.

Now we are faced with a profound contradiction in modern industrial biotechnology: the claims made for it are that it will enhance life, improve nutrition, increase biodiversity, and save the environment. Yet it seeks to achieve this through the violent manipulation of the very foundations of life.

If we look at the medical field, we see the application of biotechnology in the form of esoteric and heroic measures to defy death, almost exclusively on an individual basis. It is not vaccines for malaria or treatments for pneumonia or cholera that get the research — it's the transgenic animal organs for xeno-transplantation.

The director of the World Health Organization recently commented that "The optimism of a relatively few years ago that these [infectious] diseases could be brought under control has led to a fatal complacency." The report that he introduced points out that diseases known for centuries are now popping up in incurable strains, many of them increasingly resistant to drugs as a result of "the uncontrolled and inappropriate use of antibiotics." Making matters worse, the report points out, are modern methods of food production, such as the use of antibiotics in meat production to promote growth, but not in sufficient amounts to kill microbes, with the result that drug-resistant bacteria are then passed through the food chain to the consumer. "In the contest for supremacy", reports the World Health Organization, "the microbes are sprinting ahead."[3]

The advocates of biotechnology are quick to claim that what they are doing is simply more of what we have been doing for millennia — making bread and wine and cheese. In a sense they are correct. Biotechnology is just another expression of the drive to control not only the world around us (nature, or Creation), but also other people. It is fully in keeping with the attitude of Francis Bacon who, more than 300 years ago, as Sandra Harding points out so lucidly, used the sexual imagery of rape and torture to describe the proper scientific attitude toward nature.[4]

Technology, or technique, is a culturally embedded approach to acting on the world. It is a set of tools, particular tools that express cultural attitudes and values, such as speed, or taking things apart to see how they work. In the case of western industrial society, it is also a technology of domination and control, over nature, people, and life itself. Biotechnology is specifically the application of technology to life forms.

If we look candidly at the practice of biotechnology, we see violent intervention in the structures of life in order to reshape it according to our goals and purposes. In this respect, it can be said that biotechnology is engaged in a form of structural adjustment, but directed by Ciba-Geigy and Monsanto rather than by the World Bank and International Monetary Fund (IMF).

---

[3] Toronto *Globe & Mail*, 5/20/96

[4] Sandra Harding, *The Science Question in Feminism*, Cornell University Press, 1986, p.116.

Applied to biology, however, structural adjustment is social engineering, and this should remind us of eugenics, that is, the deliberate selection of "superior" genes, organisms, people. What else can the constant talk about "improved" seeds and crops mean? Structural adjustment, whether in the lab of Plant Genetics System (PGS) or the board room of the IMF, is about selecting preferred structural characteristics and effectively disposing of those not considered desirable.

The purpose of structural adjustment is, after all, control. And in a market economy society this control is translated into profit and the accumulation of wealth.

In biotechnology and genetic engineering, then, we have simply taken what may be the final step in a logical process. We are now engaged in the redesign of life itself, with wealth and eternal life as the goals. (We might also describe this as an expression of the fear of death.[5]) We cannot, however, honestly say that we are trying "to enhance life" in general. It is only those with the money to buy hopes of immortality that are of interest to the market.

Look at current developments in genetic testing for cancer. While many social, ethical, and medical questions remain unanswered, including questions about the efficacy and interpretation of testing itself, companies are gearing up for large-scale testing for BRCA-1 and the more recently discovered BRCA-2, genes that "cause" or predispose to cancer. "Profits are expected to be huge. The cost of Myriad's new test for BRCA-1 is going to be near $1000 a patient, and this doesn't include the cost of the doctor's examination and counseling after the results come in," according to an executive at Myriad Genetics. Myriad is itself gearing up to test 100,000 women per year.[6]

Speed kills. This has been a popular slogan to reduce automobile accidents. As in the case of the V-chip or the labeling of genetically engineered food, the slogan blames the victim, as it were, rather than putting the blame where it belongs in the first place, on the automobile manufacturers who build automobiles to go fast and sell them on the basis of performance, performance defined as acceleration and speed, or on the investors in the production of violence for TV.

Isn't it exactly the same with biotechnology? One of the industry's favorite justifications for biotechnology is the speed with which changes can be made and results achieved — on the assumption, of course, that this is inherently good. Speed is, after all, what comes with progress: faster aircraft, faster cars,

---

[5] "Having directed so many life trends in North America for the past couple of decades, baby boomers are looking to dictate new terms for death, including its elimination. With their late first marriages, late parenthood, second careers, fitness fixations and pharmacological inroads on aging, many of them — and the social institutions that cater to them — seek to defy and deny the outrage of finite existence." (Toronto *Globe & Mail*, 5/18/96).

[6] Toronto *Globe & Mail*, 7/5/96.

faster trains, faster communications of all sorts — and I just love e-mail! (Let's just ignore, for now, the contradiction between speeding fruit by jet from Chile to Canada while it has also been designed for longer shelf-life.)

Currently the biotech industry is making a great push to have its notion of the benefit of speed applied to the regulatory system. Quick approval has somehow become a desirable goal, a 'good thing'. 'You can't stand in the way of progress', we are told.

The Canadian Government's proposed revisions to the Environmental Protection Act are a good illustration of this. In its position paper the government stresses that it is "the new paradigm of global competitiveness" and "the ability to innovate and respond to new technologies [that] determines corporate success," and says that it "wants to ensure that we have a regulatory regime in place which . . . places Canadians at a competitive advantage."[7] What little it says about social or ecological health is strictly an addendum to competitiveness.

I do not think that a cynical response is out of order. If the regulation of biotechnology is subject to such mindless criteria, we will indeed be able to say that 'speed kills' — but it will be too late. The point of a regulatory process is not speeding new products to market in the name of global competitiveness. It is sanity and health — not the health of the market or the corporation, but the health of the society and all life.

Unfortunately, industry seems to see itself as above and beyond society, and its corporate health as achievable without public benefit or control. As a result, we see industry engaged in blackmail, in Germany and other jurisdictions around the world, saying that if the regulatory process is not liberalized and speeded up, it will take its business elsewhere. So now we are in a game of competing globally to see who can get away with the lowest standards. It is genuinely frightening!

The second characteristic of biotechnology is supposedly the precision of its processes. The industry makes much of this precision, although many practitioners deny it exists. Even if the precision is real, we must still question its value; after all, it is precisely the lack of precision that makes it possible for organisms to evolve and survive.

What is more interesting, however, in terms of the structural issues we should be addressing, is the fact that both speed and precision are characteristics of process, not product. So we are faced with this contradiction: the industry claims it is the process which makes the products of genetic engineering superior to traditional plant breeding, then turns around and says that as far as regulation and labeling are concerned, it is only the product that counts.

---

[7] *CEPA Review: The Government Response*, Ottawa, 1995, p.5 & p.51.

Let's look more closely at this process-product issue, because it is present at every level, from *Codex Alimentarius* to the City of Toronto's Food Policy Council.

I think the biotechnology industry is right to resist the labeling of foods as products of biotechnology. They know the public concern, and they know that the public does not make the process/product distinction that the industry would like them to. Why should the public make this distinction? In Ontario, for example, kosher dairy products have been on the market for quite a while, labeled as such. And halal is also recognized. Yet when a very substantial group of organic diary farmers wanted to market certified organic dairy products, the Ontario Milk Marketing Board fought them all the way, saying they could not make an exception to the rule of monoculture. The OMMB lost, and organic milk is now on the Ontario market and doing very well. The underlying concern here was well expressed by the exasperated comment of a western beef rancher: "What am I, inorganic?" It is assumed that the organic label indicates a superior product. This is acceptable when the group for whom the product is superior is small or marginal (Jews, Muslims, people allergic to peanuts), but it becomes terrifying when the assumed superiority has a more general appeal as a more wholesome, less high-tech or manipulated food.

The recent meeting of the labeling working group of *Codex Alimentarius* in Ottawa had to wrestle with this process/product issue: on its agenda was the labeling of halal, kosher, organic, and biotechnology foods. All are process issues, but not much headway was made on the biotech issue because there was no agreement to proceed on the basis of the U.S. position that how a food is created, grown or processed is irrelevant — at least in the case of biotech foods.

In the FDA interview cited earlier, in response to the question, "What values will these genetically engineered plants have — more nutrients, better taste?" the FDA spokesperson had an honest response: "Right now, it's more a matter of giving fruits and vegetables better shelf-life and shipping properties." In food production, genetic engineering is being applied primarily to the industrial commodity crops: tomatoes, corn, oilseeds, soybeans, potatoes and cotton. These are not crops that are in short supply or that need, as the industry calls it, improving. And what has been done to them is to make them more amenable to monoculture industrial production — regardless of what the companies might say about nutrition and the environment. These crops are being adjusted structurally in order to provide greater sales and profits for a very limited number of very large transnational corporations — corporations that are getting fewer in number and larger every day. (The merger of Ciba-Geigy and Sandoz into Novartis is just the latest and largest. Monsanto, meanwhile, is busy with another kind of expansion, having taken control of Gargiulo and Calgene in recent months.)

Even the World Bank is getting concerned. An agricultural specialist with the bank recently warned a Saskatoon audience that biotechnology might start avoiding research of public value, but little private profit. "Who is going to make the public good investments in fundamental research to agriculture?" Alex McAlla asked.[8]

At the same time, there is another sort of structural adjustment going on. One begins to realize that 'jobless recovery' and 'downsizing' and a whole lot of other euphemisms are really telling us that the corporation is no longer interested in production. It is not sales that count, as a look at the way *Forbes'* magazine rates companies reveals. What counts is equity and return on equity, dividends, and increases in share value. These are all expressions of what is now being heralded as "shareholder value". Another expression of this is the preoccupation of business with the financial markets, that is, those markets where the trading is not done in real commodities, things that people can actually eat or use to keep warm, but in invisible or imaginary commodities, such as derivatives that are based on the movement of stock indices or currency exchange rates.

If we are honest about it, we must conclude that, as presently practiced, biotechnology is going to offer nothing to the subsistence farmer anywhere — except perhaps lower prices for what little surplus he or she may try to sell. What it promises is greater 'shareholder value' for wealthy northern investors, much of this to be derived from the value-added activities of everyone from the biotechnologist to the further-processor. But this shareholder value is simply an extraction or transfer of wealth that has nothing to do with social benefit, personal well-being, or even nutrition.[9]

In the case of food, value-added is used to describe anything that might provide an opportunity to increase the selling price of a product, from simply washing potatoes instead of selling them covered with mud to turning them into pre-cooked frozen french fries, or what's worse, reprocessed, cooked, and canned "Pringles" (now P&G's biggest export product). And in practice, value-added largely means nutrition-subtracted.

---

[8] Working Paper:23/5/96 .

[9] "Imperial Oil aims to boost shareholder value" was a recent newspaper headline: "Imperial Oil is considering a number of moves to boost shareholder value, including buying back a large chunk of stock or paying a special dividend... Investors believe the company's managers will take bold steps to enhance the value of its stock. A large buyout would have that effect, as it would dramatically increase demand for the stock. At March 31, Imperial, Canada's largest integrated oil company, had $2 billion in cash and marketable securities, with $1.8 billion in debt. In 1995 it earned $514 million on revenue of $9.4 billion." (*Globe and Mail*: 5/14/96) .

When I butchered the lambs we had raised, I did not consider that I was adding value. I was, first of all, taking a life. Second, I was putting that animal in a more attractive and convenient form for human consumption. I was not adding value to the meat of that lamb as a delicious and nutritious food.

Another example from my experience as a sheep farmer. We found it necessary, for our survival, to organize a cooperative. It was a non-share capital, farmer owned and operated co-op that took charge of marketing lambs for the sheep farmers of Nova Scotia on a voluntary basis. That is, there was no legal compulsion or authority in what we were doing. What we did, in fact, was gain control of lamb marketing — from the farmer through delivery to the supermarkets. We had to concentrate power and control in order to gain a reasonable return on our labor. Not to make a profit, not to make a return on our capital, and not to increase shareholder value for the co-op, but to hopefully make a living wage.

The application of biotechnology to food is often described as 'adding value'. I can see the added cost aspect of it, particularly given the new emphasis on intellectual property rights, though the speed of new product development is outrunning the patent process. However, I have yet to see any indication of the added nutritional value that the industry uses as a selling point, particularly with gullible health professionals and their organizations, such as the American Dietetics Association (ADA) and Canadian Dietetics Association (CDA).

So when we look at agriculture, and at that application of biotechnology to agriculture, we readily see that it is not about feeding the hungry of the world, or even the growing population of the world, in fact. Nor is it even, really, about feeding the growing appetites of the growing global middle class. What it is about is making more money for corporations out of what is already being produced. We are, in fact, in the midst of a massive structural adjustment for the simple purpose of increasing shareholder value.

We are fond of referring to this new economy as an information economy, and noting that DNA is merely a form of information, which can be manipulated for profit just like other forms of information. Considering the global economic structural adjustment I have been describing, I think it may be more accurate to describe this as a post-production economy. I leave you with that thought.

Protein Production in Transgenic Animals

JULIAN COOPER
PPL Therapeutics, Inc.
Blacksburg, Va.

The agricultural industry is experiencing a period of dramatic change as new developments in biotechnology provide different methods for making products. How the industry responds to the new challenges will, to a large extent, be governed by the attitudes of the consumer. One of the major effects of biotechnology will be the altering of consumers' attitudes towards farm crops. Future farms will produce food more efficiently, perhaps in environments where originally it was considered too difficult. In addition, farms will produce many different raw materials for industrial processes, a variety of pharmaceutical proteins for human and animal use, and serve as a source of vaccines for many of the developing nations.

One of the technologies that will contribute to this expansion in new products is the generation of transgenic species, both plant and animal. The purpose of this paper is to introduce transgenic animal technology by describing what it is, how it is performed, and why it has such potential. A few examples of possible products will be discussed, and the concept of livestock "stem cells" will be introduced.

WHAT IS TRANSGENIC TECHNOLOGY?

A transgenic animal has been defined as an animal that is altered by the introduction of recombinant DNA through human intervention. The first transgenic mice were produced in the early 1980s, but it is only within the past seven years that transgenic livestock have been produced on a routine basis (Wall, R. J. 1996).

PPL Therapeutics is the world's leading producer of transgenic animals. We are based on three continents (Roslin, Scotland; Blacksburg, USA; Whakaru, New Zealand) and have the ability to produce transgenic mice, rabbits, sheep, pigs, and cows. The different species allow PPL to tailor the production route to the product's market volume and the customer's requirements for speed to market (see Table 1).

TABLE 1: CHARACTERISTICS OF DIFFERENT SPECIES

Milestones	Cow	Sheep/Goat	Pig	Rabbit
G0 birth (1)	9	5	4	1
Induced milk (1)	16	9-10	–	4
Volume of induced milk (2)	~1- >100	~1-50	–	0.01-1e
G0 adult (1)	21	13	11	6
G1 birth (1) (natural milk)	31	18	15	7
Milk Volume (3) per lactation.	>10 000	250-700	>100	~1

(1) all times are in months and relate to the starting point of microinjection
(2) all volumes are in liters and can vary considerably between individual animals
(3) all volumes are in liters
e estimate

Each species has its own quirks and difficulties, but essentially, the process is the same. A DNA construct is designed and built to express the desired protein in the animal. The site of expression is determined by the regulatory sequences used in the construct to control the coding sequence. This construct must then be introduced into a single cell embryo to allow incorporation of the transgene into the animal's genome. There are several methods available for this purpose, including retroviral transmission, stem cell transfection, and microinjection into the pronucleus or cytoplasm. The most popular and most successful method involves microinjecting the DNA solution into the pronucleus of the embryo using a very fine glass needle. The injected zygote is then transferred into a hormonally prepared recipient and brought to term. Finally, positive transgenic animals are matured and the level of expression of the transgene is determined (Palmiter et al., 1982).

DNA construct preparation

The DNA (a gene) that codes for a suitable therapeutic protein is cloned and sequenced. Every complete gene that codes for a protein has a control element (a promoter) attached to it. These direct the expression of the gene to specific sites in the body, at specific levels and times, depending on the function of the protein. PPL attaches a milk gene promoter (ovine b-lactoglobulin) to the therapeutic protein coding region to direct expression of the gene specifically to the mammary gland of the animal. In this way, the transgenic protein is produced in the milk of the animal and can be harvested and purified with no adverse effects on the animal. An added advantage is that the protein can be purified using regular dairy techniques, as well as high technology chromatographic procedures.

Transgenic production

After introduction of the construct DNA into the one cell embryo, the construct DNA is incorporated into the DNA of the cell by an, as yet, unknown mechanism, and only in a few cases are transgenic animals produced. The rate of transgenesis can be 5-25 percent of live births; it is generally believed that the larger the animal the lower the rate, although in recent experiments with cows PPL has achieved high rates of transgenesis.

The embryos are cultured *in vitro* for various lengths of time before being transferred back into a hormonally prepared recipient female. The pregnant animals are brought to term, the offspring are tested for transgenesis, and the positive ones are bred for milk production. This whole process of injection to milking takes 13 weeks in mice, 44 weeks in sheep, and 135 weeks in cows, although with the development of hormonally induced premature lactation, information on the transgenic protein's expression level and quality can be obtained within 70 weeks for the cow. Because the quantity of induced milk produced by a cow can be far greater than a natural lactation from sheep or goats (see Table 1), a lot of the purification and pre-clinical studies may be performed without having to wait for the cow's natural lactation to begin.

Why use transgenic technology?

There are various methods of protein production available commercially, including bacterial, insect, fungal, and mammalian cell culture systems. However, they each suffer, to varying degrees, from two major limitations. First, many therapeutic proteins have specific configurations that are necessary for activity. In addition, they frequently require quite complex post-translational modifications (PTM).

Some proteins require glycosylation (the addition of certain sugar residues) for activity or to ensure that they are not cleared from the circulation too quickly (Jenkins et al., 1996). Bacterial expression systems cannot perform most of these modifications and can subtly alter the folding of the protein. Yeasts and higher plants can make many of the simpler additions, but are limited in their ability to perform complex modifications.

One form of PTM found only in mammalian cells is gamma carboxylation, which is essential for the vitamin K dependent proteins, such as Factor IX, used to treat hemophilia B, and protein C, an anti-coagulant. Although these proteins can be produced in mammalian tissue culture systems, the levels of expression are very low and attempts to increase it have not been successful. Another source is fractionation from human blood plasma, but the quantities available cannot satisfy the market, and possible contamination with human viruses (HIV, or hepatitis B and C) is a significant risk.

Large complex proteins, such as fibrinogen, cannot be produced at commercial levels in any culture system available today. Fibrinogen is an hexameric molecule consisting of two sets of three polypeptides encoded by different genes, and held together by 29 disulphide bridges. For proteins such as this, transgenic animals probably represent the only possible method of production.

The second limitation is that the cost of building and running modern production facilities for cell culture is extremely high. The initial capital investment in a very expensive production facility for a product that has not been through clinical trials, and which may fail, is hard to justify. This gives two compelling reasons for the development of transgenic technology:

Production of complex proteins: The mammary gland is able to perform most PTMs that are needed for therapeutic proteins. Even a protein as large and as complex as fibrinogen can be produced in the mammary gland in gram/liter quantities.

Cost of manufacture: The initial capital investment to produce transgenic animals is one to two orders of magnitude lower than for large production facilities. Operational costs are also considerably lower.

Alpha-1-Antitrypsin

One of the most advanced transgenic products is alpha-1-antitrypsin (AAT), a serine protease inhibitor. The protein is being developed by PPL for the treatment of lung disorders, including cystic fibrosis.

The transgenic sheep line that was chosen for this product has been studied in great detail. Expression levels and transgene copy numbers have been followed over five generations of sheep, over four complete lactations in individual sheep, and between more than seven siblings, all without showing any significant change. All of the production animals are New Zealand sheep, especially flown to Scotland to avoid any possible problems with scrapie, a spongiform encephalopathy of sheep that is not found in New Zealand.

There are now nearly 150 milking ewes in the production flock, producing AAT in the new pilot production facility, the first of its kind in the world. This facility has been designed and built by PPL to combine a high-tech dairy, operating within Good Agricultural Practice (GAP), with a state-of-the-art protein purification plant operating in accordance with Good Manufacturing Practice (GMP) requirements.

Regulatory Issues

A very important part of any new technology is developing procedures and practices to ensure that regulatory concerns are addressed. PPL has been in contact with the regulatory authorities in Europe and the United States for several years to help to establish the highest levels of safety possible. The Food and Drug Administration's (FDA) Points to Consider (Aug. 22, 1995) and the European Community's CPMP Draft Guidelines (July 4,1995) on production of therapeutics in transgenic animals were generally well received by the industry. Using those documents and our own expertise, PPL has developed a regulatory strategy for transgenic animals in the production plant. That strategy begins with checking the sequence of the transgene to ensure there are no known oncogenic sequences present, and continues after the birth of transgenic animals with:

Check Insert: An analysis of copy number, transgene integrity, number of insertional sites, and stability of the insert is carried out.

Seed stock: Semen of low generation male from the same line. This could be compared to the working cell bank of a tissue culture system.

Production flock: As the seed stock comes from low generation males, the production flock will consist of several groups of half sisters with the same integration site.

Highest possible quality animals: PPL's sheep are imported from New Zealand specifically because the health status of these animals is the best in the world.

Quarantine animals: The sheep on our farms are kept in strict quarantine from other animals to minimize the risk of infection.

Control of feed stock: There is strict control over the feed stock and only approved suppliers are used.

Regular health screening: The animals are regularly checked by qualified veterinary surgeons. Animals showing any signs of disease are immediately removed from the milking flock.

Dual site production: In the future there will be several sites of production. This is a precaution in the unlikely event of a flock or herd infection at one site.

Potential Products

Protein C

Protein C is an anti-coagulant that plays an important role in the blood clotting cascade. To be active, the polypeptide must be cleaved into light and heavy chains, N-glycosylated, correctly folded and linked with disulphide bridges, and have nine gamma-carboxyl groups attached. Clearly this is a complex protein. PPL has produced fully active transgenic protein C at levels of 0.3g/l in sheep.

In addition, in blood plasma it can be seen that not all of the polypeptide is cleaved into two chains. Some single chain material remains that cannot be activated. In collaboration with Zymogenetics of Seattle, Wash., we have altered the cleavage site to increase the efficiency of cleavage to 100 percent (Foster et al., 1990). Therefore, all of the transgenic material can potentially be processed to make active protein C.

Fibrinogen

Fibrinogen is a complex plasma glycoprotein required for the final phase of blood coagulation. It is produced in liver parenchymal cells where the six chains are assembled and linked by 29 inter-chain and intra-chain disulphide bonds. Complete molecules are then secreted into the blood stream. Fibrinogen is going to be useful as a therapeutic protein and is one of the main components of surgical tissue sealants. These are being developed as a treatment for wounds as well as surgical procedures. PPL has demonstrated the production of g/l quantities of mature hexameric fibrinogen, which was functional in clotting reactions in the milk of transgenic mice (Prunkard et al., 1996), and also in induced milk from prepubertal sheep.

Neutraceuticals

In the past year, PPL established the technology to produce transgenic cattle, with nine transgenic animals born so far. Another 130 pregnant cows carrying potential transgenics have been generated. Those embryos were injected with a gene to produce a therapeutic food.

The future of transgenics

The present method of producing transgenic livestock animals, random integration of injected DNA into the genome, has two major disadvantages. First, it is inefficient and relatively more expensive than it could be. Second, currently it is only possible to add genes to an animal with no control over the site of integration. Livestock "stem cells" would address both these problems.

Embryonic stem (ES) cells are cells derived from the inner cell mass of blastocysts, which have been adapted to grow in *in vitro* conditions in the laboratory. The cells maintain their totipotency and when they are reintroduced into another blastocyst they contribute to the animal, and in some cases, to the germ line. Therefore, any changes introduced into the ES cells in culture can become part of a new line of animals. Because the cells can be propagated *in vitro,* it allows various techniques to be employed to alter the genome of the animals into which those cells will ultimately develop. For example, homologous recombination and targeting are very powerful techniques used in genetic manipulation, which have allowed researchers to mutate, delete, and/or add genes in mice (Fässler et al., 1995).

At present, totipotent ES cells can only be isolated using standard approaches from specific mice strains. It has previously been suggested by others that it may never be possible to use this type of cell for other species. However, alternate techniques are currently being investigated and we have recently achieved the derivation of totipotent ES cells from strains of mice previously shown to be intransigent to classical techniques (McWhir et al., 1996). The application of this technique to different species may allow the future derivation of totipotent ES cells from livestock. That and other approaches are directed towards allowing the transfer of genetic manipulations performed on cell lines into live animals. In some cases, these scientific approaches involve the generation of cells that could not be considered in the classic sense as totipotent ES cells. However, they may allow genetic manipulation of the genome *in vitro* and the subsequent transfer of modifications to the genome of a live animal. Therefore, when people talk of livestock "stem cells," they are usually referring to a method of production, rather than the actual cells used in the process.

Recently, Campbell et al. (1996) demonstrated that a sheep could be produced by nuclear transfer from a cell grown in tissue culture. This is the first time a route from *in vitro* cultured cells to live animals has been demonstrated in a livestock species, and offers the same opportunities for analysis and modification of gene function as are available in the mouse using ES cells.

In practical terms, it will enable transgenics to move on to the next phase in its development; giving precise control over many more useful genetic alterations in livestock animals and increasing the efficiency of production. Clearly, this new approach to transgenic production will raise ethical issues that will need to be addressed in the next few years. One of the most controversial will probably be the use of this technology for the production of transgenic animals for xenografts (Jones 1996). This is the proposed replacement of diseased human organs with those from a transgenic animal, preferably a pig. Several companies are working towards this end already, however, the application of homologous recombination and targeting would undoubtedly improve their ability to realize their goals.

It is important that this debate is conducted in public, not just in academic circles, in a rational and sensible manner, so that the ultimate consumers can have properly balanced information upon which to base their decisions.

REFERENCES

Campbell, K. H. S., J. McWhir, W. A. Ritchie, and I. Wilmut. 1996. Sheep cloned by nuclear transfer from a cultured cell line. *Nature*. 380, 64-66

Carver, A. S., M. A. Dalrymple, G. Wright, D. S. Cottom, D. B. Reeves, and Y. H. Gibson, et al. 1993. Transgenic livestock as bioreactors: stable expression of human alpha-1-antitrypsin by a flock of sheep. *Bio/Technology*. 11, 1263-1270

CBER (FDA), Points to consider in the manufacture and testing of therapeutic products for human use derived from transgenic animals. Docket no. 95D-0131, 22nd August, 1995

European Commission CPMP, Use of transgenic animals in the manufacture of biological medicinal products for human use. 4th July, 1995

Fässler, R., K. Martin, E. Forsberg, T. Litzenburger, and A. Iglesias. 1995. Knockout mice: How to make them and why. The immunological approach. *Int. Arch. Allergy Immunology*. 106, 323-331

Foster, D. C., C. A. Sprecher, R. D. Holly, J. E. Gambee, K. M. Walker, and A. A. Kumar. 1990. Endoproteolytic processing of the dibasic cleavage site in the human protein C precursor in transfected mammalian cells: Effects of sequence alterations on efficiency of cleavage. *Biochemistry*. 29, 347-354

Jenkins, N., R. B. Parekh, and D. C. James. 1996. Getting the glycosylation right: Implications for the biotechnology industry. *Nature Biotechnology*. 14, 975-981

Jones, I. 1996. 2010 - a pig odyssey. *Nature Biotechnology*. 14, 698-700

McWhir, J., A. E. Schnieke, R. Ansell, H. Wallace, A. Colman, A. R. Scott and A. J. Kind. 1996. Selective ablation of differentiated cells permits isolation of embryonic stem cell lines from murine embryos with a non-permissive genetic background. In Press

Palmiter, R. D., R. L. Brinster, R. E. Hammer, M. E. Trumbauer, M. G. Rosenfeld, N. C. Birnberg *et al.* 1982. Dramatic growth of mice that develop from eggs microinjected with metallothionein-growth hormone fusion genes. *Nature*. 300, 611-615

Prunkard, D., I. Cottingham, I. Garner, S. Bruce, M. Dalrymple, G. Lasser, P. Bishop and D. Foster. 1996. High-level expression of recombinant human fibrinogen in the milk of transgenic mice. *Nature Biotechnology*. 14, 867-871

Wall, R.J. 1996. Modification of milk composition in transgenic animals. In: *Beltsville Symposia in Agricultural Research, XX, Biotechnology's role in the genetic improvement of farm animals*. Eds. R. H. Miller, V. G. Pursel and H. D. Norman. Savoy, Ill.: American Society of Animal Science

Phytoremediation

BURT ENSLEY
Phytotech, Inc.
Monmouth Junction, N.J.

Phytoremediation is a new application of biotechnology — an exploitation of specific plants to clean up the environment. Plants are used to treat or remove environmental contaminants from soil and other solids like dredge spoils, water, other liquids, and air.

Plants may be viewed as the original polluters. Our earth was hospitable to methanogenic bacteria before the evolution of photosynthesis by plants, producing oxygen that now is 20 percent of our atmosphere. Oxygen forced the methanogens into anaerobic regions of the world. Now we plan to use plants to help us clean up toxic soils. Scientists at the Ag Biotech Center at Rutgers University, such as Ilya Raskin, have explored methods of using plants to treat organic and inorganic contaminants in soils, and have discovered that plants provide very effective biological processes for cleaning up and removing contaminants from the environment.

I will focus on the use of plants to remove and concentrate heavy metals. Heavy metal contaminants in the soil and water in all developed countries are a large and important worldwide problem caused by manufacturing processes and other activities of commerce dating back to the 1700s. One may view a plant as a solar driven concentrator of chemicals like metals that took 360 million years to build, test, and optimize. The scientists at the Ag Biotech Center enhanced that process so it is more rapid and useful for clean up at contaminated sites.

What is the appeal of phytoremediation? First, it is an elegant method of treating contaminated sites relative to conventional treatment methods in which contaminated soil is transferred to a landfill (which means simply moving it somewhere else), presumably for permanent storage, and then transferring uncontaminated soil to the original site. Plants leave the soil in place while removing the contamination. Phytoremediation is cost effective, while transfer of contaminated soils to a licensed hazardous waste site is expensive.

Even after 25 years of cleanup efforts and growing awareness of the dangers posed to the environment by contaminated sites, America still has a huge number of contaminated sites. The main reason for the slowness of the cleanup is the large expense of using conventional technology to clean up these sites. We must find more cost effective methods. The use of plants helps address the economic issue. This use offers owners a new method of remediating contaminated property at less cost, which presumably will accelerate the pace of remediation.

Phytoremediation is compatible with public concerns about technology. People have been concerned about biotechnology. They are frightened particularly about genetically engineered bacteria. They always will be. It is easy to frighten people about bacteria. For example, when you tell people how many bacteria there are in yogurt, it doesn't make them think better of the bacteria, it makes them quit eating yogurt. Plants are something that everyone can see; they are familiar and therefore comfortable with plants. Using plants to clean up a contaminated site is more acceptable to community members than the use of bacteria.

At the time of the Exxon Valdez oil spill, the state of Alaska decreed that no non-Alaskan micro-organisms could be used. This was the first microbial-immigration policy and it was useless. Bacteria travel on the wind, and in many other ways, and they didn't seek Alaskan regulators' permission before they entered Alaskan waters to eat the oil on the beach.

Phytoremediation is sophisticated scientifically. Technology must be developed so it can be safe, effective, and reliable.

Over the next five years the United States will spend $43 million to treat only sites contaminated with heavy metals and heavy metals mixed with organics. That's not what would be needed to clean up all the sites. We need low-cost approaches so that we can accelerate the pace of remediation.

Phytoextraction is the subset of phytoremediation in which plants remove contaminating toxic or radioactive metals from a site by root uptake and accumulation in the plant shoots.

I will describe the use of a plant — Indian Mustard — to remove lead from a toxic site. The plants are watered with agricultural chemicals that make the lead water soluble. The water containing lead is taken up by the roots and moved into the stems and leaves where it evaporates, leaving the concentrated lead in the plant. The harvested plants may be sent to a smelter where the metal is recycled — a hazardous waste is removed from the environment and recycled to commerce. This technology, developed by the Ag Biotech Center, is the subject of a technology transfer agreement between Rutgers University and Phytotech.

Brassica juncea (Indian Mustard) has the special property of metal accumulation that was discovered at the AgBiotech Center by screening plants for their metal uptake abilities. This metal uptake property is being expanded and enhanced by research including genetic engineering, mutagenesis, and selection at the Ag Biotech Center and at Phytotech. Most plants do not accumulate heavy metals in the stems or leaves and are not useful for phytoextraction. For most plants, some metal is taken up by the roots, but not much is transfered to the shoots and stems. Indian Mustard and Amaranthis accumulate cadmium, nickel, zinc, copper, lead, and chrome6, a widespread pollutant in New Jersey and a carcinogen.

To increase availability of lead, a biodegradable, short-lived chemical is added to the soil to cause the plants to take up the lead. An example is the metal chelator EDTA added to contaminated soil. There is a direct correlation between the amount of EDTA added and the metal uptake. We are also doing environmental impact rates and effect studies to evaluate the environmental impact, good and bad, of removing metal from the soil by growing the plants on site and adding chemicals to induce uptake and clean the site.

We are in the process of modeling, analyzing, and monitoring the impact of this approach to remediating contaminated sites. What do you do with the harvested plants following phytoextraction? We dry the plants, may incinerate them, may compost them, or may compact them by pressing followed by acid leaching and disposal, or recycling, of the heavy metal. The volume or mass of the plants that contain the extracted heavy metals is only about two percent of the volume or mass of contaminated soil. This is a 50-fold concentration.

Indian Mustard has many attractive characteristics for cleaning up toxic metal sites. It is inexpensive to grow and grows very rapidly to about five feet in about two months, producing 2-10 tons/hectare of dry weight with fairly rapid clean up of the site. We believe there are a significant number of contaminated heavy metal sites in the U.S. that could be remediated in about three years or less so that the Environmental Protection Agency or the state environmental agency would conclude that that site was safe for use. We have great hope that we will be able to use phytoremediation as an effective technology.

We have treated a site in the middle of an urban environment, Magic Marker, in Trenton, N.J. It was originally contaminated by a battery manufacturing and recycling facility located there since about 1950.

Other sites that we have examined in Trenton and other New Jersey towns were contaminated in the 1700s and 1800s by practices such as pottery making. Trenton was a big pottery making center at one time. Pottery glazes contained lead or cadmium and there is a large number of lead and cadmium contaminated sites in the city of Trenton.

The Magic Market site was blocked out, samples were taken to establish lead content, the site was plowed, and fertilizer was added. The soil on the site is dark, not because of humus, but due to the asphalt in the soil. The site was once a parking lot. An irrigation system was installed and Indian Mustard was hydroseeded to evenly distribute the seeds without disturbing the soil.

We collaborated with the local residents. There was a Phytoremediation Day in Trenton. Through an organization called Isles Inc., we attended community meetings, told everybody that lived in the area (there are houses all around this site) what we were doing, and invited everyone to attend the first planting. The costs of those trials were paid by Phytotech, which had raised $6 million of venture capital and has 16 employees.

As indicated in my presentation, we are enthusiastic about the role of phytoremediation in the cleanup of toxic sites.

Agricultural Biotechnology: A Farmer's Perspective

Mary Lou Garr
Ontario Federation of Agriculture
Beamsville, Ontario, Canada

I am particularly pleased and somewhat surprised to be a speaker today, given that I am neither a scientist nor an expert in biotechnology. Frankly, I have only a limited understanding of biotechnology. I'm just a plain ordinary family farmer from a long line of plain ordinary family farmers. However, this humble status has never prevented me from striving to make as much money as possible and I see exciting opportunities ahead to do just that: rapid improvements in the quality of crops, livestock, and livestock products; huge reductions in input usage and costs of production; and the creation of new consumer products. In fact, I expect that my family farming operation will be fundamentally changed over the next few years because of the things most of you are doing.

However, in the push to advance the field of agricultural biotechnology, farmers are often the last to be considered or consulted. Does the research always meet our production needs? Does the regulatory system provide for timely approvals of new technologies? Are farmers sought out to participate in forums like this?

At NABC 7, held in May, 1995, I met very few farmers, though I found our input in workshops much appreciated. This year, I doubt I'd see many hands raised if I asked how many people rely on farming for their income. Yet farmers are your critical link in the biotechnology path to market. We utilize the research and produce the product which then must be marketed and find public acceptance. And all this in a world where the anti-technology crowd is way out in front in shaping public opinion.

Most farmers are cautiously optimistic about agricultural biotechnology, but they need someone to take leadership in managing this controversial issue on their behalf.

Because the theme of this conference proposes partnerships, I would like to describe chronologically the slow and often tortuous journey of my own general farm organization in its attempt to foster new partnerships to manage the agricultural biotechnology issue.

First let me tell you about the Ontario Federation of Agriculture (OFA). The OFA has a membership comprising 40,000 Ontario farm families and 28 organizations, mostly commodity specific. Its mission is to improve the economic and social well-being of farmers. The responsibility of fulfilling that mission rests with a seven-person executive committee that is governed by a board of 100 elected directors. I am one of those directors. So who would be better placed to attempt to draw together the broad range of stakeholder groups necessary to manage the agricultural biotechnology issue? It should have been easy, right? Not quite!

1993

In June 1993, the OFA hosted a workshop on the Impact of Biotechnology on Agricultural Production and invited agricultural leaders from across the province to attend. It attracted a disappointingly low 30 participants, but was the first attempt by the OFA to address agricultural biotechnology issues and, I believe, was the first attempt to do so by any Ontario farm organization.

That initial workshop had only two objectives. The first was to have speakers provide updates on biotechnology as a production tool in both plant and animal agriculture, regulatory aspects of biotechnology, and ethical questions associated with biotechnology and agriculture (not a subject any of us wished to tackle).

The second objective was to have workshop participants discuss their views on the impact of biotechnology on agricultural production, on our ability to compete globally, and our responsibility to feed a hungry world. Break-out groups were asked to consider three questions: 1. What did you hear today that concerns you?, 2. How should the agricultural industry respond to these concerns?, and 3. What is the next step?

The concern most often expressed was about the lack of information on agricultural biotechnology. It was observed that both producers and consumers need to be better educated with regard to this technology. One participant suggested that there was a huge gap between reality and perception that needed to be filled. Another individual questioned whether or not the production tools resulting from agricultural biotechnology would be universally available to farmers — all farmers, whether large or small in scale. It was felt that more workshops were needed, but a broader range of stakeholders should be invited. For the first time, the notion of forming a coalition of organizations having an interest in agricultural biotechnology was raised. We had taken our first tentative steps into deep and controversial waters.

1994

In May 1994, the Chair of the OFA's Environmental Committee, who also served as an OFA Vice President, attended NABC 6 at Michigan State University along with an Ontario contingent of farmers, professors from the University of Guelph, and farm organization staff. The objective was to become familiar with the conference theme, Agricultural Biotechnology and the Public Good, and draw on the material to develop an agricultural biotechnology position for Ontario's agricultural industry.

Then things became slightly complicated. A position paper planned for distribution at the OFA's annual meeting was never written. Even worse, the OFA Vice President, who had taken the most interest in addressing agricultural biotechnology issues, was not re-elected. Consequently, by December 1994, the OFA found itself in the unfortunate position of not having moved forward on the direction provided by the June 1993 workshop. We were in danger of losing our momentum just when farmers were experiencing frustration and controversy over the BST battle, and Bt corn and potatoes were quickly coming to market.

1995

The OFA had a much more productive year in 1995. In January, our Environment Committee, of which I am a member, nominated Paul Verkley, a registrant here at NABC 8, as chairman and took the initiative to identify agricultural biotechnology as one of our primary areas of emphasis.

In May 1995, four members of the OFA Environment Committee and one OFA staff member traveled to NABC 7 in Columbia, Missouri. The Committee made a conscious decision to utilize resources in this area, given their commitment to keeping abreast of agricultural biotechnology issues. While at the conference, it became apparent that the OFA was the only farm organization in North America that gave agricultural biotechnology this level of priority.

However, I would be less than honest if I did not report to you that some members of the OFA Executive Committee strongly disagreed with the Environment Committee's decision to send a significant delegation to NABC 7. Ostensibly, their objection was because of the cost, but I think it really demonstrates the subtle conflict within the farm community on biotechnology, and the real difficulty in developing a unified position.

Then in the fall of 1995, farm groups began to make progress. The OFA arranged a meeting where representatives from farm organizations, agricultural input suppliers, food processors and retailers, consumer associations, government (provincial and federal), and the University of Guelph conversed on the subject of agricultural biotechnology. The OFA was delighted that 50 people attended this meeting, remembering that there were only 30 people in attendance at our meeting in June 1993.

Part of the agenda was to ask all registrants to briefly outline the position on agricultural biotechnology held by their organization, agency, or institution. While that exercise clearly demonstrated the wide variance of opinion both between and within groups, it led to an agreement that some coordination was needed. From that meeting came a proposal for an Ontario Agricultural Biotechnology Committee that should have three distinct roles: communication, advocacy, and consensus building.

While there was some discussion as to who should take the lead role on this committee, it was quickly decided that the OFA, as a general farm organization with no particular vested interest, was best suited. By the time the meeting adjourned we had a long list of stakeholder organizations to be contacted and invited to participate in the inaugural meeting. These included some groups seldom asked to directly partner with farm groups, such as agricultural input suppliers, grocery distributors, consumer advocates, academics, and government bureaucrats.

1996

As of June 1996, the Ontario Agricultural Biotechnology Committee has met four times and, in my view, has been enormously successful in bringing to the same table a number of disparate players in the agri-food industry to discuss an extremely controversial topic. In the short time the committee has been together, it has established a mission statement, a set of goals, and a subcommittee structure that breaks issues into three categories: communication, research, and regulation.

The committee's mission is to foster the growth and development of agricultural biotechnology for the maximum benefit of the Ontario agri-food sector and the community-at-large.

The stated goals are:

1. Instill a higher level of knowledge and understanding of agricultural biotechnology within the agricultural industry.
2. Effectively communicate within the agricultural community, and between the agricultural community and society-at-large.
3. Influence future biotechnology research and commercialization.
4. Encourage a timely, science-based assessment of biotechnology products for potential use in the Ontario marketplace.

Establishing this committee was certainly an accomplishment, but not without some difficulties. For example, some within the farm community question whether we should be focusing our communications efforts on educating farmers about agricultural biotechnology. In their view the emphasis should be directed only to consumer education. There have also been questions regarding the role of the OFA, or more specifically, the appropriateness of having the OFA's Environment Committee take the lead in managing agricultural biotechnology issues. And within the OFA itself, there is still considerable debate as to where agricultural biotechnology fits into their existing committee structure. But I ask, where would the issue fit any better than in the Environment Committee? Despite these minor irritants, I can tell you that the Ontario Agricultural Biotechnology Steering Committee is generally supported and acknowledged as the best vehicle for developing unified agri-food sector positions on issues relating to agricultural biotechnology. And as a farmer, I want to see that happen soon.

From the perspective of farmers, having unified positions on agricultural biotechnology issues is absolutely critical. It is ironic that the rate of scientific discovery in the field of agricultural biotechnology is advancing far more rapidly than is our ability to address the issues which are raised as a result of these discoveries. It has taken the OFA a considerable length of time to get to the point where we have now assembled a committee capable of initiating a process whereby agricultural biotechnology issues can be managed to the satisfaction of the agri-food sector. I, for one, look forward to the committee accomplishing its mission of fostering the growth and development of biotechnology for the maximum benefit of the Ontario agri-food sector and the community-at-large. As a farmer, I can't afford to have it fail.

PART V

LIST OF PARTICIPANTS

Ainsworth, Sharon A.
State Relations 1140
Administration 325
Rutgers University, College Avenue
New Brunswick, NJ 08903

Adrich, Jane
University of Wisconsin-Madison
1205 Warf Bld
610 Walnut Street.
Madison, WI 53705

Arntzen, Charles J.
Boyce Thompson Institute
Tower Road
Ithaca, NY 14853

Baker Segelken, Jane
NABC
419 Boyce Thompson Institute
Tower Road
Ithaca, NY 14853

Baldy, Marian
School of Agriculture
California State University
Chico, CA 95929-0310

Barefoot, Susan
Clemson University
Biological Sciences Dept.
A203J P&AS Building
Clemson, SC 29634

Barton, Kenneth A.
Agracetus, Inc.
8520 University Green
Middleton, WI 53562

Baumgardt, Bill R.
Purdue University
AGAD Building, Room 116
West Lafayette, IN 47907

Browning, C. B.
Oklahoma State University
Agricultural Division
139 Agricultural Hall
Stillwater, OK 74078-6019

Burke, Michael J.
Oregon State University
College of Agricultural Sciences
Corvallis, Oregon 97331

Byrne, Barbara
Natural Science & Mathematics
Richard Stockton College
Pomona, NJ 08240

Chase Jr., Theodore
Rutgers University
Dept. of Biochemistry & Microbiology
Lipman Hall
Cook College
New Brunswick, NJ 08903

Chess, Caron
Ctr. for Environ. Communication
Rutgers University, Cook College
George Road Annex
New Brunswick, NJ 08903

Chin, Chee-Kok
Rutgers University, Cook College
Dept of Plant Science
263 Foran Hall
New Brunswick, NJ 08903

Cohen, Spencer
Rohne Poulence Ag Company
P. O. Box 12014
Research Triangle Park, NC 27709

Cooper, Julian
PPL Therapeutics, Inc.
1700 Kaft Drive
Blacksburg, VA 24060

D'Itri, Frank M.
Michigan Ag Experiment Station
Michigan State University
109 AG Hall
East Lansing, MI 48824

Dahl, Dennis A.
1301 Broadmoor Drive
Champaign, IL

Day, Peter
Ctr. for Ag Molecular Biology
Rutgers University, Cook College
Foran Hall 204A
New Brunswick, NJ 08903

DeFuria, Dianne
Bristol-Meyers Squibb Company
P. O. Box 4000
Princeton, NJ 08543-4000

Decter, Stephen A.
Ecopolicy Center
Rutgers University, Cook College
Martin Hall 109A
New Brunswick, NJ 08903

DeWitt, Jerry
Iowa State University
2104 Agronomy Hall
Ames, IA 50011

Dickinson, William P.
VA Dept. Of Agriculture & Consumer
1100 Bank Street Suite 210
Richmond, VA 23218

Dobert, Raymond C.
Biotechnology Information Center
NAL/USDA
10301 Baltimore Ave. 4th Floor
Beltsville, MD 20705-2351

Doremus, Jeanne
484 Sheppards Mill Road
Bridgeton, NJ 08320

Doyle, Dennis
Tru-Blu Cooperative
New Lisbon, NJ

Dunn, Allen
Clemson University
Natural & Recreational Resources
102 Barre Hall
Clemson, SC 29634

Ensley, Burt, CEO
Phytotech, Inc.
1 Deer Park Drive, Suite 1
Monmouth Junction, NJ 08852

Evans, David
DNA Plant Technology
6701 San Pablo Avenue
Oakland, CA 94608

Evans, Kenny
Arizona Farm Bureau
4141 El Dorado Road
Phoenix, AZ 85040-1625

Fields, Fredrick O.
Nutritional Regulatory Affairs
Wyeth-Ayerst Laboratories
145 King of Prussia Road A-2
Radnor, PA 19087

Furey, Peter J.
New Jersey Farm Bureau
168 West State Street
Trenton, NJ 08608

Gain, Jeff
AARC Center/USDA
P. O. Box 607
Hardin, IL 62047

Garr, Mary Lou
Ontario Federation of Agriculture
417 Thirty Road
RR2 Beamsville, Ontario
Canada LOR 1B2

Germida, Jim
University of Saskatchewan
College of Agriculture
5 Campus Drive
Saskatoon, SK S7N 5A8

Ghai, Geetha
Rutgers University, Cook College
CAFT, Room 221-A
New Brunswick NJ 08903

Giamporcaro, David E.
US EPA
401 M Street SW
Mail Code 7405
Washington, DC 20460

Goldburg, Rebecca
Environmental Defense Fund
25 Park Avenue South
New York, NY 10010

Gray, J. Ian
Michigan Ag Experiment Station
Michigan State University
109 AG Hall
East Lansing, MI 48824

Groote, Joyce
Food Biotechnology Communication
90 Spark Street
Suite 1127
Ottawa, Ontario
Canada KIP 5B4

Hance, B.J.
Ctr. for Environ. Communication
Rutgers University, Cook College
Georges Road Annex
P. O. Box 231
New Brunswick, NJ 08903

Hardy, Ralph
NABC
419 Boyce Thompson Institute
Tower Road
Ithaca, NY 14853

Harvey, Bryan
University of Saskatchewan
Horticulture Science Dept.
51 Campus Drive
Saskatoon, SASK Canada S7N5A8

Henricksen Karen
University of Nebraska-Lincoln
Center for Biotechnology
N300 Beadle Center-19th & Vine St.
Lincoln, NE 68588-0665

Jones, Richard L.
University of Florida
1022 McCarty Hall
Gainesville, FL 32611

Kehrig, Ron
Ag-West Biotech, Inc.
230-111 Research Drive
Saskatoon, Saskatchewan
Canada S7N3R2

Klessig, Daniel F.
Rutgers University
P. O. Box 759
Piscataway, NJ 08855

Kneen, Brewster
The Ram's Horn
32463 Beaver Drive
Mission, BC V2V 5R3 Canada

Kraus, Martine
University of California - Berkeley
Ctr. For Study of Law & Society
2240 Piedmont Avenue #2150
Berkeley, CA 94720

Krueger, Charles R.
Penn State College of Agriculture
217 Agricultural Administration Bldg.
University Park, PA 16802

Landry, Laurie G.
Boyce Thompson Institute
Tower Road
Ithaca, NY 14853

Lacy, William B.
Cornell Cooperative Extension
Cornell University
276 Roberts Hall
Ithaca, NY 148-53

Lawton, Michael
Rutgers University, Cook College
Foran Hall
New Brunswick, NJ 08903

Laskin, Allen
Laskin-Lawrence Associates
383 S. Middlebush Road
Somerset, NJ 08873

Lee, Tong-Ching
Rutgers University
P. O. Box 231
New Brunswick, NJ 08903

Lee, George E.
Ag Research
204 Kirk Hall
Univ. Of Saskatchewan 117 Science P
Saskatoon, SK S7N 5C8

Lohr, John, Asst. Dir
Utah State University
Biotechnology Center
Logan, UT 84322

Lund, Daryl B.
Agriculture & Life Sciences
Cornell University
260 Roberts Hall
Ithaca, NY 14853

MacGillis, Miriam Sister
Genesis Farm
41 A Silver Lake Road
Blairstown, NJ 08725

Manghi, Stephen
Cook College
Continuing Professional Education
101 Ryders Lane
New Brunswick, NJ 08903

Marcotte, Bill
Clemson University
Biological Sciences Dept.
132 Long Hall
Clemson, SC 29634

Martin, Marshall A.
Purdue University
Dept. Of Agricultural Economics
Krannert Building
West Lafayette, IN 47907-1145

Meahger, Laura
Rutgers University, Cook College
Martin Hall
New Brunswick, NJ 08903

Mitchell, Roger
University of Missouri
2-69 Ag. Bldg.
Columnia, MO 65200

Mroczko, Jeanne
NJ DEP
CN 423
401 E. State Street
Trenton, NJ 08625

Mullner, Stefan
Hoechst AG
65926 Frankfurt am Main
Germany

Myers, Oval
Southern Illinois University At Carbondale
Professional Plant Breeding
Dept. Of Plant & Soil Sciences
Carbondale, IL 62901

Olson, Ken
AFBF
225 Touhy Avenue
Park Ridge, IL 60068

Porteous, Ken
University of Guelph
RR 1
Simcor
Ontario, Canada N3Y439

Rauh, Bradley
University of Missouri–Columbia
308 Curtis Hall
Columbia, MO 65211

Ripton, John
School on the Environment
747 Backhus Estates Road
Glen Gardner, NJ 08826

Rouder, Clifford
Long Island University
CW Post Campus
Box 394
Greenvale, NY 11548

Sander, Eugene G.
University of Arizona
PO Box 2100036
Tucson, AZ 87521-0036

Schroeder, Joseph J.
Michigan State University
208 G.M. Trout FSHN Bldg.
East Lansing, MI 48824-1224

Slifer, Lynn K.
1 Schick Road
Milford, NJ 08848

Snow, Judy
Rutgers University, Cook College
Ag BioTech Center
Foran Hall 204C
Trenton, NJ 08625
New Brunswick, NJ 08903

Swan, Patricia B.
Research/Advanced Studies
Iowa State University
211 Beardshear Hall
Ames, IA 50011

Thompson, Paul B
Texas A&M University
Ctr. for Biotech. Policy & Ethics
College Station, Texas 77843

Thornton, David
University of Waterloo
27 Munson Cres
Scarborough, Ontario
Canada M1P3M

Tillotson, Lucia
Middlesex County College
155 Mill Road
Edison, NJ 08837

Underwood, Allen
Helena Chemicals
6075 Popler Avenue
Suite 500East 5th Floor
Memphis, TN 38819

Verkley, Paul
Ontario Federation of Agriculture
40 Eglinton Avenue
Toronto, Ontario
Canada M4P3B1

Walter, Rick
Canadian Institute of Biotechnology
130 Albert Street Suite 420
Ottawa, Ontario
Canada K1P5G4

Watkins, Lynne H.
American Cyanamid Company
APRD
PO Box 400
Princeton, NJ 08542

Weeks, Donald P.
University of Nebraska - Lincoln
Center for Biotechnology
N300 Beadle Center
Lincoln, NB

NOTES

NOTES